# THE
# PERPETUAL
# NOW

ALSO BY MICHAEL D. LEMONICK

*The Light at the Edge of the Universe*
*Other Worlds*
*Echo of the Big Bang*
*The Georgian Star*
*Global Weirdness* (primary author)
*Mirror Earth*

# THE PERPETUAL NOW

*A Story of Amnesia, Memory, and Love*

Michael D. Lemonick

DOUBLEDAY

NEW YORK LONDON TORONTO

SYDNEY AUCKLAND

 Published in the United States by Doubleday, a division of Penguin Random House LLC, New York, and distributed in Canada by Random House of Canada, a division of Penguin Random House Canada Limited, Toronto.

www.doubleday.com



*Book design by Pei Loi Koay*
*Jacket design by Emily Mahon*

Library of Congress Cataloging-in-Publication Data
Names: Lemonick, Michael D., 1953–
Title: The perpetual now : a story of amnesia, memory, and love / Michael D. Lemonick.
Description: New York : Doubleday, [2016]
Identifiers: LCCN 2016017750 | ISBN 9780385539661 (hardcover) | ISBN 9780385539678 (ebook)
Subjects: LCSH: Johnson, L. S. (Lonni Sue)—Health. | Encephalitis—Patients—Biography. | Brain damage—Patients—Biography. | BISAC: SCIENCE / Life Sciences / Neuroscience. | MEDICAL / Neuroscience. | BIOGRAPHY & AUTOBIOGRAPHY / Medical.
Classification: LCC RC390.L46 2016 | DDC 616.8/320092 [B] — dc23
LC record available at https://lccn.loc.gov/2016017750

MANUFACTURED IN THE UNITED STATES OF AMERICA

10 9 8 7 6 5 4 3 2 1

First Edition

*In memory of Margaret Kennard Johnson*

# CONTENTS

# PREFACE

One day a couple of years ago, a woman about my own age approached me on the street in Princeton, New Jersey, where I live. "I'm Aline Johnson," she said. "I'm not sure whether you remember me. Have you heard about what happened to my sister?"

No, I hadn't heard, but I did remember Aline vividly. Even before she spoke, I knew exactly who she was. In her late fifties at the time, Aline had thick, wavy brown hair that fell to her shoulders. An image immediately came into my mind of this same face, the same hair, the same voice, but in my memory she was thirteen years old, dragging a cello case into the music room at the Valley Road middle school in Princeton, where we both played in the orchestra back in the 1960s. I could hear Mrs. Switten, the director, calling out Aline's name—not just the words, but also the woman's reedy, northern midwestern–accented voice as she spoke them. When she got really exasperated at one of her students, she'd say, "Jeeminy Christmas!"

That led immediately to another memory: the sinking feeling I would get in the pit of my stomach before every school assembly. Mrs. Switten insisted I play a bugle call on my trumpet as they brought in the American flag. I flubbed it every time. My classmates were unkind about it, every time. Some of them were unkind about my clothes, too, and my haircut, as I recall, which were both very uncool.

Simply seeing Aline Johnson's face had instantly transported me back to middle school. Her reference to her sister, meanwhile, triggered a different set of memories. I remembered the sister's name, for one, with no prompting: it was Lonni Sue. I knew that Lonni Sue and I had been in high school together, and that she was several years older. I remembered how peculiar her name had seemed to me, better suited to someone living in the backwoods of Kentucky than to a girl raised in an Ivy League college town. I remembered that Lonni Sue had played an instrument, as well, though I didn't know which one. I vaguely remembered that she'd gone on to become some sort of artist. I couldn't picture Lonni Sue's face, since we'd never interacted. The only visual image that came to my mind was that odd name printed on a high school concert program. In the string section somewhere, I thought.

This rush of memories—visual, auditory, emotional—lasted just a fraction of a second, launched by the sight of a woman I'd encountered no more than once or twice since the Nixon administration. I barely knew her, even in middle school. We'd never had a single conversation, even though we were just one grade apart and had orchestra together several times a week. But somewhere in my brain, these different scraps of information had been sitting, waiting for some event to cause a small group of neurons—that is, brain cells—to fire with a simultaneous burst of electrochemical energy.

I experienced that crackling of cellular activity as a series of sights, sounds, and feelings, brought immediately back to my consciousness a half century after these details first took hold.

Memories can be so vivid that the neuroscientific explanation for what they actually are—merely a discharge of energy, the synchronous firing of brain cells in patterns laid down minutes or hours or decades in the past—threatens to trivialize their power as we experience them, to downplay the essential role they play in our lives. Walking into a familiar room, or seeing a familiar face, or hearing a voice we know well triggers a rush of context, reminding us of who we are in relation to this person or place or object. Memories allow us to navigate the world—literally, but also cognitively and emotionally. Without a personal history to call on—without access to the events and people and experiences we've had over a lifetime—it's hard to imagine having any sort of identity.

If we have no memories of the experiences that made us, how can we know who we are?

That's why I was so appalled when Aline Johnson told me, as we stood on a street corner in Princeton that day, what had happened to her sister. Several years earlier, a virulent brain infection had wiped Lonni Sue's memory nearly clean. She could no longer remember her past, except in vague generalities. She could no longer form new memories that she'd be able to rely on in the future, except in the most rudimentary way. Try to imagine what that might be like: a friend walks up to you—not a distant acquaintance, like Aline was for me, but someone you've shared your life with for many years—and you have no idea who she is. She tells you her name. Doesn't ring a bell. She reminds you that you saw a movie together just the other day, and describes the actors and the plot. Nothing. She brings up the man who laughed so hard he spilled his popcorn all over your lap. You have no clue. That's a trivial example, but now imagine that it happens every time you meet this person, and that you don't forget just the silly stuff. You also forget that this person helped get you through your divorce, and that she stood by you when your father got cancer and died a long, painful death. You don't even remember that you were divorced, or that you were ever married in the first place. You don't know that your father died. If someone reminds you, you feel terrible grief, because your memory loss isn't so complete that you've forgotten your immediate family. But you forget again right away, so that if someone reminds you of his death the next day—or even ten minutes later—you grieve again, as though for the first time.

It isn't just this friend that you can't recognize or recall. It's pretty much every friend you ever made, and every memory you ever made with him or with her. You can't recall your first kiss, or your first love, or your favorite vacation, or the teacher who inspired you, or the one who gave you a hard time in calculus. For most of us, walking down the street where we grew up or stepping into the apartment where we lived during some particularly exhilarating or challenging time in our lives brings back a flood of impressions so powerful it feels as though we're reliving them. For Lonni Sue, visiting most of those places brought back . . . pretty much nothing at all. It's like having Alzheimer's disease, but worse: with Alzheimer's, aside from the very last stages, you retain memories of the distant past. You remember your childhood, often vividly. Lonni Sue didn't even have that.

It wasn't just the past that Lonni Sue had lost, however.

. . .

In the typical Hollywood depiction of amnesia, the victim can't look backward. That was true of Lonni Sue as well. But she also couldn't look forward. It might not be immediately obvious, but thinking about the future—about what might happen tomorrow or next week or next year, and planning for what you might do, calls on memory. What do I enjoy doing? What things do I need to accomplish? Who haven't I seen for a while? Without knowing what I've done in the past, or who my friends are, or what the options are, there's no way to imagine the future. Memory is so central to learning, identity, purpose, decision-making, and relating in a meaningful way to others that it's fundamental to who we are. Lonni Sue had lost an essential part of herself, and her doctors were convinced it would never return. What that actually meant for Lonni Sue as a person, however, wasn't at all clear to me as Aline and I stood there talking. What could it be like to live without memory? What must it be like to be inside her head, to live your life in a perpetual "now"? If I were in this situation, I thought, I'd probably end up sitting in a dark room, profoundly depressed. How did Lonni Sue handle it?

Aline suggested I find out for myself. A few weeks later, I rang the front doorbell at the house Aline and Lonni Sue had grown up in, in a residential section of Princeton. The door opened directly into the dining room. Sitting at the head of the table, with papers strewn all around her, was a plump, attractive woman in her early sixties, with a round face and chin-length reddish hair, held in place by a headband. She was wearing a white turtleneck and a black fleece vest, and as I stepped through the door, she looked to see who had come in. Her face lit up with an enormous smile, as though she couldn't imagine a nicer surprise. "Hello!" she said, her voice unexpectedly warm and rich and welcoming. "My name is Lonni Sue," she said. "What's yours?" I told her. Then she asked, "Would you like to see my drawings?"

The table was littered with them—sheets of paper decorated with figures of horses, cats, fruits, stars, suns, moons, and letters of the alphabet. She began pointing the images out one by one, then looked up at me. "Have you ever sung the alphabet?" she asked. "It's really fun!" This woman clearly wasn't depressed. She seemed childlike in her openness and enthusiasm, but the drawings were exquisitely executed, and the alphabet song she proceeded to demonstrate was something no child could have pulled off. This wasn't the "A-B-C-D" song we all learned as kids. It was something, Aline later said, that Lonni Sue had come up with herself. She would start with a word beginning with the letter A, and sing her way through the alphabet, one word per letter, all the way to Z. The song was always precisely twenty-six words long. Both the tune and the words were improvised every time, and usually formed a sentence of sorts.

Here is what she sang for me that morning: "Artists beautifully creating delightful exquisite finery giving hospitable inspiration joining keen laughter's monthly necessities openly preparing quiet refreshment sweetly turning under violet weathervane xylophones yearning zestfully." It didn't make a lot of sense, but she performed it without hesitation (although she did stretch out a word once or twice as she reached for the next).

By the time she was done singing the song, she'd forgotten that she wanted me to try it. This was the first obvious indication that something wasn't quite right. The second indication came a few minutes later, after she'd excused herself to use the bathroom. When she returned to the dining room, Lonni Sue saw me standing there waiting for her. "Hello!" she said, brightly, smiling that same brilliant smile. "My name is Lonni Sue. What's yours?"

# THE PERPETUAL NOW

# INTRODUCTION

The temperature is hovering at around zero degrees Fahrenheit in rural Cherry Valley, in upstate New York, when a dairy farmer named Buzz Stetson finishes up the predawn milking and walks a few hundred yards to knock on Lonni Sue Johnson's door. It's December 30, the day before New Year's Eve, in 2007. A series of snowstorms has rolled through over the past couple of weeks, burying the hilly farmland under several feet of snow that won't melt until sometime in April.

Buzz comes over to visit pretty much every morning. Lonni Sue isn't just his neighbor; she is also his landlord, his business partner, and his friend. If Buzz could have things his way, they would be lovers as well, but Lonni Sue isn't interested. She's made this firmly but gently clear. It isn't in her nature to be unkind. Buzz's unrequited passion makes things a little complicated, but it doesn't prevent them from sharing a cup of tea, or from sitting in her small farmhouse kitchen on frigid mornings or warm summer afternoons when the flowers she's planted on every side of the house are in full bloom—hollyhocks, lilies, bee balm, irises, butterfly bushes. She has a huge lilac that puts out deep purple blossoms in spring, and there are raspberry and blueberry bushes as well. Lonni Sue has a collection of cats thirteen deep, Buzz likes to say—Squawk, Schnitzel, Mouse, Tippy, Tomato, Squeak, Babe, and more. During the warm months, they swarm visitors on the porch. When someone shows up, they come running from every corner of the house to assemble in the kitchen.

On a normal day, Lonni Sue and Buzz might sit and talk about the organic dairy business they started together, or about the neighbors. He's lived there much longer than she has, and he went out of his way to introduce her around when she moved in. Or they might talk about a day trip they're planning, to look at a new piece of farm machinery, or to take an organic farming workshop together. Sometimes they'll talk about their lives before they came to know each other—about what brought each of them to the place she calls Watercolor Farm, where she sleeps in the house and he lives in a trailer out on the property.

In December 2007, Buzz and Lonni Sue are both in their late fifties, so they're well matched in age. Otherwise, they make a very odd couple. He's about six feet tall, bearded and gaunt, invariably dressed in torn, dirty jeans and a plaid flannel shirt covered in stains. He was raised on a farm in western Massachusetts, went to college for a while, then dropped out to go back to work on his father's farm. "I never really wanted book knowledge," Buzz likes to say. He's fiercely intelligent, but in a rough-hewn way that matches how he looks and dresses.

Lonni Sue, by contrast, has a wholesome, round face. She isn't beautiful—you'd be more likely to call her pretty, or cute—but she has a dazzling smile. When you talk to her, she listens with an intensity and focus that makes you feel as though you're the only person in the world. She has a shy warmth and compassion that invariably draws people to her. "She's easy to fall in love with," one of her closest friends says. Lonni Sue grew up in a privileged university town. Her father was an extremely successful electrical engineer, her mother an equally successful sculptor and printmaker who, among other things, taught classes for decades at the Museum of Modern Art (MoMA), in New York.

Lonni Sue is an artist as well. She's drawn covers for *The New Yorker* magazine, illustrated dozens of books, and done work for the *New York Times* and a long list of corporate clients. She's also an accomplished amateur violist and a private pilot who owns not one but two small airplanes. She writes a column for the local newspaper, reminiscent of the dispatches the *New Yorker* staff writer E. B. White used to write from his farm up in Maine. She lived for nearly two decades in New York City when she was in her thirties and forties, hanging out with some of the most successful artists, musicians, writers, and academics of her generation.

And yet here she is, on an isolated farm about ten miles to the northeast of Cooperstown, New York. The place is beautiful—a mix of fields and woodlands, sloping upward to a high point with a spectacular view of Otsego Lake, which lies on the other side of the hill. It's a place where she can think. "It was like meditation for her to be there," her mother would tell me later. The reason she moved so far from the city, and from her family, was that she wanted a place where she could bulldoze her own landing strip, so she could be able to walk out of her house and just take off. At least, that was the story she and her family usually told.

On the freezing morning of December 29, Buzz is knocking on Lonni Sue's door, not just for a cup of tea, but to invite her to a party—his brother's birthday party, over in Vermont. It's the sort of thing she might ordinarily have done, but this time she says no, in a way that he finds a bit unsettling. "She wasn't acting right," he said.

"She was vague, drifty, you know what I mean? She wasn't herself. She said, 'You just go.'" So he goes to the party. It's a four-hour drive each way. By the time Buzz gets back and finishes his chores, it's well after dark. He doesn't see Lonni Sue at all that night.

The next morning, he knocks again to see how she's feeling, but she doesn't answer. That seems odd, he thinks. Her car is out front, so he knows she's home. He goes around to look through the window of her home office. There's Lonni Sue, sitting bent forward, with her head on her desk, examining the computer mouse. She looks bewildered. It's as though she'd never seen such a thing before. Buzz taps at the window. She doesn't respond. So he goes around to the basement door, puts his shoulder to it, and breaks in.

When he finally gets her to focus, Lonni Sue insists she's just fine. Clearly, she isn't. She's confused, and she keeps acting bizarrely—digging in her flowerpots with a fork, for example. Buzz thinks she might have had a stroke, but he really has no idea. So he calls Kay Anichini, a neighbor who works at Bassett Hospital, down in Cooperstown. Kay is a pharmacy technician, not a doctor, but she knows more about medicine than he does. Maybe she'll be able to figure out what's going on. By the time Kay gets to the house, Lonni Sue's speech is slurred. She's no longer making any sense at all. "We told her she had to go to the emergency room," Kay said. "She didn't want to go." In the end, they convince her to get into Kay's car by telling her she can wear her favorite pilot's hat. "It was like she was a child," Kay recalls.

They leave for the hospital at about five that afternoon. Kay's daughter, Maya, who is fourteen at the time and who knows and admires Lonni Sue, insists on coming along. Buzz has to stay behind. Another bad snowstorm is coming, and if he goes with them he might not be able to get back in time to milk the cows. "That stretch of road up there is miserable in the winter," Kay said later. "It blows like there's no tomorrow."

By the time they reach Bassett, Lonni Sue's condition has deteriorated badly. She no longer recognizes either Kay or Maya. She's running a dangerously high fever. It isn't immediately clear to the doctors what's going on, since confusion and fever can be a sign of many different illnesses. One of those tends to loom large in the minds of emergency-room physicians, however, because while it's relatively rare, it can have horrific consequences.

*Viral encephalitis* is the generic term for an acute inflammation of the brain caused by one of several viruses. (You can also get encephalitis from bacteria or parasites.) Some of these viruses are transmitted by mosquitoes—West Nile virus and eastern equine encephalitis virus, for example. But another is a microbe that most of us already have in our bodies, usually without being aware of it. It's called *herpes simplex virus 1,* or HSV1. Its most common symptom is cold sores. (Genital herpes is usually caused by a related virus, HSV2.) Many people who are infected with HSV1 never show any symptoms at all. Others have intermittent outbreaks of sores that last a week or two. In between outbreaks, the virus spends most of its time in a state of latency, or inactivity, lurking in nerve cells until something wakes it up—generally, doctors believe, something that puts stress on the body or the immune system, such as an allergic reaction, a bacterial or viral infection, physical trauma, psychological stress, a bad sunburn, or even menstruation.

Usually, reactivating HSV1 just means another round of cold sores. In rare cases, however—only a few hundred in the United States each year, according to the Centers for Disease Control and Prevention—the awakened virus begins inching its way up the nervous system, via the olfactory nerve or the trigeminal nerve, and into the brain. Nobody has been able to explain why HSV1 decides to do this. Although the physicians at Bassett weren't sure at first, that's what had happened to Lonni Sue.

Once HSV1 reaches the brain, it can burn through gray matter rapidly, consuming neurons like an out-of-control wildfire tearing through dry brush. It's not clear how much of the resulting damage is caused by the virus itself and how much comes from friendly fire by the body's immune system trying to fight it off. In any case, up to

30 percent of people infected with HSV encephalitis die, even with treatment. Those who survive can suffer permanent memory loss or other mental impairments.

Because it's so devastating, HSV encephalitis is always at the back of doctors' minds, despite its rarity. If there's even a chance that a patient is infected, explained Deborah Sentochnik, chief of infectious diseases at Bassett Hospital, physicians assume it is HSV1 encephalitis until they can prove otherwise. (Privacy laws dictate that Sentochnik isn't allowed to say whether she treated Lonni Sue personally, although it's not unreasonable to suspect that she did. Anything she says on the topic for quotation, however, has to be generic.)

The standard treatment for HSV1 encephalitis is a drug called acyclovir, delivered intravenously. So while the doctors awaited definitive lab results from blood tests and a spinal tap, Lonni Sue got the drug, just in case. She had no idea what was happening to her. She tried to pull out her IV lines, so the nurses had to strap her to the bed. Kay and Maya stayed with her through the night, trying to keep her calm.

While acyclovir can kill HSV1, it can't repair any of the damage the virus has already done. That's why doctors administer it even before they're sure what they're dealing with. The downward spiral from an initial headache and mild fever into acute, life-threatening illness can occur over just a few days, but since confusion is one of the symptoms, the victim usually isn't aware of what's happening. If Lonni Sue hadn't lived alone, someone might have noticed that she was behaving oddly and dragged her to the hospital. If Buzz had somehow connected her driftiness the day before with a possible brain infection, she might have received her first dose of acyclovir a potentially crucial thirty-six hours earlier than she actually did. If Lonni Sue's mother and sister had been up from New Jersey to spend the holidays with her, as they'd often done in the past, they might have noted her behavior even earlier.

That's not how it happened. In the end, the treatment saved

Lonni Sue Johnson's life. Before the virus was defeated, however, it had destroyed billions upon billions of brain cells. The damage wasn't random. Another thing doctors don't understand about HSV1 encephalitis is why it preferentially attacks neurons in the brain's medial temporal lobes, the twin regions at the core of the brain, one on each side, where our everyday experiences are converted into long-term memory.

The hippocampus in particular is crucial to this process. The hippocampus is a vaguely seahorse-shaped structure, nestled deep within the brain ("hippocampus" means seahorse in ancient Greek). Every normal brain has two hippocampi, one in each hemisphere. It is here that sights, sounds, smells, tastes, and sensations, along with the thoughts those experiences trigger and the emotions they arouse, are linked together to form a coherent memory that might be retrieved sometime in the future. In some ways, the hippocampus is like the dutiful host at a party who introduces the guests to one another then steps back to let them form a direct relationship.

The guests—the collection of experiences you're having at any given moment—are then linked, in other parts of the brain, to experiences from the past, forming a network of associated memories, like the ones that seeing Aline Johnson called up for me. Some of these are enormously rich because they have been repeated many times and have a lot of emotional content. Your mother's face, coupled with her voice, with her quirks and traits, with how you feel about her, and with the thousands of different situations and interactions you've shared with her, might be an example. You've thought of her a million times, and it doesn't take much to bring her to mind again. Other memories are less likely to come up spontaneously. I messed up the bugle call in front of my peers a dozen times, maybe, and I don't remember any specific instance. But it happened often enough and was mortifying enough that I can remember what it generally felt like, as long as something reminds me of it. Others are highly specific—for example, exactly where I was and how I felt and what happened next when my father called to say he thought my mother had had a stroke. The more eventful or emotional an experience is, or the more often it's repeated, the easier it is to recall, but completely trivial information can be dredged up from memory as well. A couple of years ago, I realized that a woman I was talking to had gone to Princeton High School. I probably didn't know her, she said, but I might have known her older brother, Charlie Wheeler. Instantly, an image of Charlie appeared in my head. We had barely interacted with each other during high school, and I literally hadn't had a single thought about him since then. But he was tucked away in there nevertheless.

Similarly, you probably can't recall what you had for lunch a week ago Wednesday unless something important happened at the lunch, or unless some cue related to that particular meal comes up to trigger it. But you can almost certainly remember your wedding or your high school graduation or where you were when JFK was shot or when 9/11 happened, years or even decades afterward.

Without at least one functioning hippocampus to stitch those sights, smells, thoughts, and feelings together, however—without a host to introduce them—you almost certainly couldn't remember even those major life events. Neuroscientists know this largely thanks to a man named Henry Molaison, whose hippocampi and surrounding tissues were surgically removed in 1953 in an attempt to control his severe epileptic seizures. The operation worked, but it had another consequence nobody had anticipated. Until that point, scientists who studied the brain believed that the hippocampus was mostly responsible for processing the sense of smell. But not long after Molaison emerged from the operation, it became clear that he had essentially lost the ability to form new memories of people and events. If you were introduced to him, then left the room and came back, he'd have no idea he'd ever met you before, no matter how many times you repeated the exercise. He had become densely amnesic, in the jargon of neuroscience.

The same thing would happen to Lonni Sue Johnson. The virus that attacked her brain inflicted damage to her hippocampus and other parts of her medial temporal lobes similar to what surgery had done to Henry Molaison. An MRI scan of her head would show a black nothingness of dead tissue at the core of her brain.

At first, when she emerged from her fevered confusion, she couldn't walk or talk or feed herself. All of those things returned in time. But her ability to turn new experiences into long-lasting memories—that essential function that allows us to learn, to construct an ongoing personal history, to record the past and anticipate the future—was almost entirely gone, never to return. If she sees a movie, she won't remember the title or the plot or the actors ten minutes after it ends. If she meets someone new, then sees them again a day later (or even five minutes later, as I discovered for myself), she'll have no idea that she ever saw them before. This first became clear to Aline and their mother, Maggi, while Lonni Sue was in the very earliest stage of recovery, just a week or two after she arrived at Bassett. As her confusion began to lift, Lonni Sue didn't seem to recognize the doctors and nurses and aides who came into her room many times every day.

Even worse, the encephalitis robbed Lonni Sue of most of the memories she had formed before the illness struck. She knew who her mother and sister were, even though she couldn't speak their names at first. But she didn't know Buzz when he came to see her. She didn't know Kay, or her daughter Maya. She didn't know her other neighbors, or any of the dozens of friends she'd made in the area. She had shared meals and projects and musical evenings and walks through the countryside with many of these people. She had confided in them, and shared her deepest, most troubling secrets with a few. Suddenly, they were complete strangers. This didn't seem to bother Lonni Sue, since she had no memory of what she was missing. For the visitors, it was baffling and disturbing.

Both forms of memory loss happened to Henry Molaison as well. Neuroscientists now know that it can take years for the hippocampus to fully consolidate experiences into permanent storage. When his hippocampi were surgically removed, the backlog of partially processed memories was cut off. You had to go back several years before his operation to find things he could recall. Even then, he couldn't bring specific episodes to mind, the kind of memories that start with "I remember the time when . . ." He could recall only general facts about his life of the sort that began "I grew up in . . ." or "I used to . . ."

It's the same with Lonni Sue. As she emerged from the more acute phase of the illness and began to speak, it gradually became clear that there were enormous gaps in her memory of the past. She doesn't know the names of her teachers in high school. She doesn't recognize a photograph of John F. Kennedy, who was assassinated when she was in eighth grade. She knows Maggi and Aline, but when you show her photographs of her aunts and uncles and cousins, she recognizes only some of them. She has no memory of her high school prom, or her college graduation—assuming she even went, which nobody seems to know—or her wedding day, or her divorce ten years later (she had no children, which now seems like a blessing). She does know that she once had two airplanes, and she reminisces about how liberating it felt to fly. Today she also understands that her father, Ed Johnson, is dead, although she's evidently not clear on whether it happened a year or a decade ago. (He died in 1989.)

Lonni Sue returned to her farm in Cherry Valley just once, a few months after her illness. She had loved this place. She'd transformed the barn into a studio, and built the organic dairy business with Buzz, and bulldozed herself a landing strip, and walked the property every night, gathering sticks she would use for kindling. But when she came back for that visit, she barely recognized any of it. It all seemed very pleasant to her, and she was perfectly cheerful, but as far as she knew, she'd never been in this place before.

These memories, along with just about every other memory she'd accumulated in fifty-seven years of life—the events and experiences and people that in many ways defined who she was—were gone forever. When you meet Lonni Sue Johnson today, she seems awake and aware and fully engaged with you. She greets you warmly and enthusiastically, and if you come back the next day, she does it again as though it were the first time she'd ever met you. For Lonni Sue, it is the first time, every time. It's completely unclear what the experience of life must be from her perspective, trapped in a perpetual "now" that will never turn into "then."

Except for two factors particular to her case, the story might have ended when Lonni Sue was finally released from a series of hospitals and rehabs, several months after the virus first attacked her brain. She could have lived the rest of her life in an apparently meaningless, never-ending present. But from the moment Lonni Sue emerged from the acute phase of her illness, her mother and her sister poured most of their physical and emotional energy into helping her recover. If Maggi and Aline hadn't followed her from hospital to rehab to nursing facility, supplementing the care of professionals by working with her day after day on her physical and mental recovery, she might still be bedridden. Without their constant attention, a computer mouse might seem like an utterly alien object to her even today.

The second thing that happened was that, purely by chance, Lonni Sue got connected to a team of neuroscientists at Johns Hopkins University. They knew, as anyone who ever took freshman psychology knows, about Henry Molaison and the key role he played in the scientific understanding of how memory works. They recognized that Lonni Sue could be even more important, given that her talents and interests and experiences were so much richer than Molaison's had been. There was far more to test with Lonni Sue, so they could probe much subtler aspects of memory than an earlier generation of scientists had done with Molaison. Not only that: the noninvasive brain scans that have transformed neuroscience over the past couple of decades—primarily function magnetic resonance imaging, or fMRI—became available only late in Molaison's life, after he'd begun to develop dementia on top of the damage he'd sustained from surgery.

As a result, while the research on Henry Molaison had laid out some basic facts about memory—that the hippocampus is crucial to conscious memory, for example, and that some forms of unconscious memory are processed elsewhere in the brain—that was only the beginning. This is often the way it is in science, where a single, transformative discovery is just the start of true understanding. When William Harvey first described the human cardiovascular system in the 1600s, for example, it was only the crudest road map to what would eventually become modern cardiology. When Copernicus figured out that the Earth orbits the Sun rather than the other way around, in the 1500s, it was just a rough framework that would guide astronomers to a true understanding of the universe—an understanding that wouldn't come for centuries. In a similar fashion, Henry Molaison's case established just the basic outlines of the anatomy of human memory.

Even before Lonni Sue lost her own memory, however, neuroscientists were beginning to realize that the hippocampus might do far more than convert experience into conscious memories for later recall. Experiments with both animals and other amnesia victims had begun to suggest that these brains-within-a-brain do nothing less than orchestrate our entire internal representation of the outside world. According to this new thinking, the hippocampus constructs mental maps of the relationships between experiences, objects, locations, people, and more. Without these maps, both conscious and unconscious, we'd be helpless at navigating the world both physically and in the sense of understanding our relationship to the present, the past, and the future.

Yet it's clear that while Lonni Sue has lost a conscious connection to the past and the future, she can still function in the present. She has access to the mental maps that tie perception to action, for example. If she reaches for a cup, she can grasp it and pick it up and drink from it without hesitation. She has access to the maps that tie concepts to words. She can read fluently. It might be that the hippocampus is more crucially important for some functions—conscious, explicit memories in particular—than for others. It might be the sort of party host who can let some guests go about their business on their own while needing to tend more carefully to others.

Neuroscientists still don't completely understand the full extent of what the hippocampus does, which is why Lonni Sue's case is so important. Before her illness, she navigated many specialized cognitive worlds—visual art, instrumental music, aviation—in addition to the ordinary ones the rest of us encounter. This gives neuroscientists a chance to test the emerging view of the hippocampus's role in ways they've never been able to do. For her mother and sister, it's reassuring to think that this might at least lead to some important scientific insights. Over the next several years, they will impress this on Lonni Sue over and over again. Despite her drastically impaired capacity to retain new information, this much will finally take hold. "Will this help other people?" she invariably asks at the start of each testing session, and again at the end. She always seems reassured when her mother or her sister or one of the neuroscientists who are studying her responds, "Absolutely."

*Chapter 1*

# A TEXTBOOK CASE

Hartford, Connecticut, is a traffic-choked city today, but in the 1930s it was a sleepy town where kids could play in the streets for hours without having to move aside for a passing car. One day in 1934, an eight-year-old boy named Henry Gustav Molaison fell off his bike and struck his head. He wasn't wearing a helmet, since it wouldn't occur to anyone for another half century that children needed that kind of protection. Or maybe it happened a little differently: maybe he was on foot and another bike knocked him over. His parents didn't remember which it was, when they thought about the incident years afterward.

Henry might have lost consciousness for a few minutes, or perhaps not. Concussions were considered a normal part of growing up back then, so nobody would have paid much attention. After the accident, whatever it was, he seemed perfectly normal. If you'd told Henry's parents at the time that their only child would one day become the most celebrated neuroscience patient in history, and that his obituary would appear on the front page of the *New York Times,* Gus and Lizzie Molaison—he was an electrician, a Cajun who came originally from Thibodaux, Louisiana; she was a stay-at-home mother whose parents had emigrated from Northern Ireland—would have been convinced you'd lost your mind.

A couple of years after he hit his head, Henry began to have epileptic seizures, and his parents wondered if the bike accident might have been responsible. At first, they were mild, petit-mal seizures. Henry would go into what seemed like a trance for a couple of minutes, then return to normal. By the time he was in his mid-teens, however, he was having grand-mal seizures, in which he would fall to the floor, go into convulsions, foam at the mouth, and bite his tongue. These episodes were so disruptive and so embarrassing that he dropped out of one high school, waited several years, then started again at another school as a seventeen-year-old freshman. He didn't take the college-prep track, choosing instead the school's vocational course. After he graduated, at twenty-one, he got a job with a company that built electric motors, then moved on to work at an assembly line, making typewriters.

Throughout his early twenties, Henry's epilepsy kept getting worse. By the time he was twenty-six, he was taking four powerful medications, any one of which by itself would have kept most patients' seizures in check. He'd had a series of tests to rule out a brain tumor or some other physical abnormality. The most unpleasant of them, called a *pneumoencephalogram,* involved draining the cerebrospinal fluid, which normally cushions the brain, then forcing air into Henry's skull to replace it, which allowed an X-ray to look for any unusual gaps in his brain tissue. That showed nothing. Neither did a much less invasive electroencephalogram, or EEG.

At a loss for how else to proceed, Henry's doctor, an eminent neurosurgeon named William Beecher Scoville, suggested an operation as a last resort. Scientists understood by this time that neurons, the cells that process and store information in the brain, communicate with one another through tiny electrical signals. They also knew that seizures are caused when these signals go out of control, sending spasms of electrochemical energy bouncing chaotically through the brain. In some cases, the chaotic signals originated in areas that had been damaged by illness or injury. Henry's parents were right to suspect that his bike accident might have been the cause of his epilepsy.

The tests found no such injuries, but epilepsy can also be caused by much subtler brain abnormalities, which are often genetic. Several relatives on Henry's father's side had had epilepsy as well, so maybe that was the explanation. No one knew where these abnormalities were located, but tests on both animals and humans had shown that deliberately stimulating the brain's medial temporal lobes with mild jolts of electricity could trigger seizures. Scoville himself had removed medial temporal tissue from patients with schizophrenia, hoping to cure their psychosis. It didn't work, but one patient who also had epileptic seizures saw significant improvement in that illness.

So Scoville proposed to remove large parts of Henry's medial temporal lobes. "This frankly experimental operation was considered justifiable," he would write several years afterward, "because the patient was totally incapacitated by his seizures and these had proven refractory to a medical approach." Scoville made it clear to Henry and his parents that the operation was risky. He thought it was likely to relieve the epilepsy, but there might be side effects. But the seizures were so awful that the Molaisons decided it was worth the risk. Even if the surgery didn't cure him, Henry said, in giving his consent, it might help others.

On August 25, 1953, orderlies wheeled Henry into an operating room at Hartford Hospital, his skull newly shaved, and the surgery began. Scoville drilled two large holes in Henry's forehead, each about an inch and a half across, and reached in with an instrument called a brain spatula, which he used to lever the frontal lobes—one on each side—out of the way. This gave him access to the hippocampus. Then he worked the spatula in and down, and began to apply suction to the soft, spongy brain tissue. By the time he was done, Scoville had sucked out big chunks of Henry's two hippocampi, one on each side of the brain. He'd also removed much of the neighboring amygdalae, which are involved in the processing of emotions. And he'd taken out tissues from the entorhinal, perirhinal, and parahippocampal cortices, which border the hippocampi. He didn't get all of the hippocampus out on either side, because the far ends curved away, out of reach of his straight, inflexible suction tube. The remaining chunks of hippocampus, however, were now largely disconnected from the rest of the brain. They might as well not exist.

Once Henry came out of the anesthesia, it quickly became clear that Scoville's invasion of his brain had been largely successful in treating the epilepsy. Henry's seizures weren't eliminated completely, but they were dramatically reduced in both frequency and intensity. It was also clear, however, that something else had happened, something awful. Henry recognized his parents. His intelligence, which was above average according to presurgical testing done in the hospital, was intact. So was his personality. He'd been shy, friendly, and good-natured before the operation. He was the same afterward.

But Henry had no clue about who the doctors, nurses, and other hospital staff who came to his room were, even after seeing them repeatedly. He could remember new information for a minute or two—if you said, "Good morning, Henry, my name is John," and then, right afterward, "What's my name?" he had no problem giving the right answer. If you asked him again fifteen minutes later, however, he couldn't do it. He couldn't navigate his way to the bathroom even after having been there several times. He didn't know why he was in the hospital. An hour after he'd had lunch, he couldn't remember eating. He'd eat a second lunch if you put it in front of him, and then a third. He couldn't remember much of anything about his past—the condition known as retrograde amnesia. And he also had anterograde amnesia: he couldn't convert his everyday experiences into new memories to be recalled in the future. Until his death fifty-five years later, he never would.

Scoville was appalled. The operation was, he would later acknowledge, a tragic mistake. He'd had no idea that this might happen. He didn't know at the time that Wilder Penfield, in Montreal, had had a similar outcome with two other patients, known only by their initials, F.C. and P.B. Both men had epilepsy, like Henry (who himself would be known simply as H.M., to protect his privacy, until after he died). Penfield had operated on them in an attempt to cure it. He'd removed medial temporal lobe tissue, including the hippocampus, from one side of their brains only. Penfield had done the same surgery on many other patients; these two were the only ones who developed profound amnesia. It turned out that, unlike the others', the mirror structures on the opposite sides of F.C.'s and P.B.'s brains weren't normal. Both men had suffered injuries to those tissues long before Penfield performed his surgery—possibly as early as birth. Penfield's other patients had backup. They could use the undamaged lobes on the other sides of their brains to take up the slack of memory formation and retrieval.

Scoville read about these two cases not long after he'd performed his surgery on Henry. He was naturally intrigued, and contacted Penfield, who suggested that a young colleague of his named Brenda Milner should come down from Montreal to test H.M., along with some of Scoville's schizophrenia patients who had also had medial temporal lobe surgery. Milner, born in England and educated at Cambridge, had been a graduate student in the laboratory of Donald Hebb, a pioneer in the attempt to understand how the brain creates and maintains memory. When Penfield began his campaign to cure epilepsy through surgery, he had recruited Milner to help him test patients to see how the operations might have affected their cognitive abilities. She had helped him figure out why F.C.'s and P.B.'s memory functions had been so severely affected, and she was eager to see how this new patient down in Connecticut might fit into the story.

The testing took place in 1955. Milner showed rigorously what Scoville had understood only in general terms: Henry's intelligence, language, and other higher mental functions were unaffected by the operation. His personality was essentially unchanged. He was described by the scientists who studied him as "gentle, goodhearted and altruistic." But his memory had been profoundly damaged, in both the past and future direction.

Unlike Scoville's schizophrenia patients, Henry wasn't mentally ill before his operation, which would have made it more difficult to disentangle memory problems caused by the surgery from any memory problems that might have been caused by the schizophrenia instead. And unlike Penfield's patients P.B. and F.C., who still had some function in the memory structures that had been damaged naturally rather than surgically, Henry had virtually none on either side of his brain. Milner and Scoville wrote up the results in a paper that appeared in the *Journal of Neurology, Neurosurgery, and Psychiatry*. It's no exaggeration to say that this article, which first established the crucial role of the medial temporal lobe in memory, changed the course of neuroscience.

As Milner and her colleagues found during later rounds of testing, Henry did retain some memories of his past, but they were almost invariably general, not specific. They were along the lines of "I remember that I went to X high school," in contrast to "I remember that crazy time in French class when Jimmy let the hamster loose and Monsieur Picard screamed *'sacre bleu!'*" The former are what neuroscientists now call *semantic memories;* the latter are *episodic memories,* since they record specific episodes in our lives rather than generalities. Henry knew where he'd grown up and where he'd gone to high school. He knew that his mother and father used to take him on vacations along the Mohawk Trail, a scenic highway that cuts across the Berkshires in western Massachusetts. But he couldn't remember a single specific event from one of those Mohawk Trail vacations, for example, or something unusual that happened in high school, or a gift he'd gotten on a particular birthday.

There were just two exceptions. In 1939, when Henry was thirteen, Gus and Lizzie bought him a ride in a small plane as a junior high school graduation present. This adventure was so profoundly exciting, it was as though it had burned itself into Henry's memory too deeply for even surgery to cut it out. Decades after the operation, he could still recall and describe the flight—what it looked like, what it sounded like, how it felt—almost moment by moment, in detail. Henry also retained a vivid memory of the first time he ever smoked a cigarette, which happened when he was ten. Presumably, it was as unpleasant as the plane ride was thrilling.

Aside from these two events, the only memories he had were semantic. That applied not just to memories of his own life, but also to his general knowledge about the world, which is also part of semantic memory. You almost certainly know that Paris is located in France, but you probably don't know when you first learned that fact. There's no specific experience, no episode you can point to where the information first entered your head. Similarly, you probably also know where your parents grew up. You know what an orange tastes like. You know that the United States fought against Germany and Japan in World War II. But you probably don't know exactly when or how you came by these bits of information.

Semantic and episodic memories are different, but they both share this feature: you can describe them in words. Other memories are impossible to articulate, since they don't involve conscious thought. Riding a bicycle, hitting a tennis ball, driving a nail, and playing a musical instrument are all things you remember how to do after you've learned, but you can't really explain what it feels like to do them. People sometimes call this sort of thing "muscle memory," because it seems as though your muscles, not your brain, are doing the remembering.

Neuroscientists now call those two broad categories of memory *declarative,* which refers to things you can describe in words, and *procedural* or *implicit,* things you simply know or know how to do without being able to describe them (retrieving the words to construct a sentence is another example). Until Henry came along, nobody really knew whether procedural and declarative memories were formed or stored or retrieved from the same part of the brain, or different parts, or how the whole process worked.

Thanks to the purity of his memory loss, however, which was untainted by any accompanying mental illness or loss of other cognitive abilities—understanding speech, for example, or recognizing objects—H.M. became what Suzanne Corkin, the MIT neuroscientist who worked under Brenda Milner and went on to study Molaison for more than four decades, calls the "gold standard for the study of amnesia . . . the yardstick against which other amnesic patients were judged." In the scientific literature, she writes, they were typically referred to as being "'as bad as H.M.' or 'not as bad as H.M.'"

Over decades of careful study, Corkin, who died in the spring of 2016, either performed or directed hundreds of research studies to probe the tiniest nuances of Henry's memory loss. Once a month or so, on average, starting in 1966 and continuing until the mid-2000s, Henry would be driven from Hartford up to MIT, where Corkin had settled after doing graduate work with Milner in Montreal. Corkin herself, one of her graduate students or postdocs, or one of dozens of independent scientists who worked with Henry over those decades would put him through a battery of tests. Many of the tests had been created just for him.

Over that time, Henry became more than just a subject. "He was a friend," Corkin told me, sitting in her MIT office one afternoon. "And not just to me. He was a member of our lab family." For the first few years after the operation, Henry continued to live with his parents—and then, when his father died, in 1966, with his mother. After she came down with dementia, he moved in with a relative and, finally, in 1980, into a nursing home near Hartford. After that, Corkin said, he had nobody to visit him or look after him. "We sort of filled in the gaps as best we could. We used to send him presents for Christmas and we'd send him pictures of our dogs and stuff like that." When a health issue came up, Corkin had the nursing home call her. "I don't do this with other patients," she said. "If something happens with them, they go and see their doctor, but with Henry, I wanted to make sure that he was well taken care of and that his needs were met, medical and otherwise." She meant this literally. The last time she visited Henry, in the fall of 2008, she asked his caregivers what he needed. White socks, they told her. So this eminent neuroscientist at one of the most prestigious universities in the world drove over to Costco, bought a pack of socks, and mailed them to Connecticut. She would never see Henry alive again.

*Chapter 2*

# "WOULD LIKE TO MEET CHARLEMAGNE"

It's September 1968. Lonni Sue Johnson and her parents are on the road, heading west. Lonni Sue is about to start her freshman year at the College of Wooster, in Ohio. The choice is not a common one for high school kids from Princeton, New Jersey. It's not that they tend to go to the famous hometown university. Princeton does take a dozen or more students from Princeton High each year, which is an unusually large number for a public high school. But that leaves hundreds of other seniors who have to find another place. They tend to spread up and down the Northeast corridor, although a few head for small, excellent liberal-arts colleges in other parts of the country. Long before *U.S. News* started its college-rankings franchise, guidance counselors at Princeton

High were touting places like Reed College, in Oregon; Grinnell, in Iowa; Beloit, in Wisconsin; and more, as "undiscovered." In the 1960s, that was actually true, at least for easterners.

The College of Wooster, in the town of Wooster, about sixty miles south of Cleveland, in northeastern Ohio, was even more undiscovered than most. Lonni Sue would probably never have heard of it except that her mother grew up in the town and went to college there. Today the school is mostly known for its focus on undergraduate scientific research. In the late 1960s, however, it was especially strong in history. Lonni Sue might instead have ended up heading for Oberlin College, also in Ohio, with its renowned music conservatory. She was such a good viola player that she was allowed to sit in with the Princeton University Orchestra while she was still in high school. In college, however, she wanted to focus on history, maybe even become a historian. Though her father was an electrical engineer, he read in pretty much every field you could imagine. He especially loved history and frequently shared what he'd been reading at dinner with the family. This was evidently inspiring to Lonni Sue. "Would like to meet Charlemagne," it says in her high school yearbook, next to a black-and-white headshot. Lonni Sue is round-faced, with a bowl haircut, wearing a dark sweater over a white shirt with Peter Pan collar—the senior-photo uniform for girls of that generation.

Ed Johnson drove the family station wagon that day, starting out very early in the morning. They crossed quickly from New Jersey into Pennsylvania, and then, after seven hours or so on the Pennsylvania Turnpike, into Ohio. It was here, somewhere between the state line and the outskirts of Wooster, that Lonni Sue abruptly announced a change of plan. She didn't want to study history after all. "I really think I want to go into art," she said. And Lonni Sue being unusually stubborn, even for a teenager, that was pretty much that.

Her father was startled by this declaration, which came entirely out of the blue. Her mother was a little distressed as well. "Here she was, on her way to a school that had once been outstanding in art, but really wasn't any longer," Maggi would tell me, more than four decades later, sitting in the living room of the same midcentury-modern house in Princeton they'd set out from on that long-ago September morning. No one knew this better than Maggi Johnson. She had grown up in Wooster, and had gone to the college herself, in the late 1930s. At the time of our conversation, Maggi was ninety-seven years old, a number that seemed hard to reconcile with the sharp-minded, energetic, active woman who sat across from me. It's true that she walked carefully, since a fall at her age could be life-threatening, and she was a bit stooped over with osteoporosis. A couple of days earlier, however, I'd run into her at the local gym, where she would drive herself to pump iron under the supervision of a personal trainer several times a week. "She's kind of amazing," the trainer said, and it was clear that this was an understatement. Maggi was also finishing up some new work for an art exhibition that would be opening in just a few weeks. She was probably too busy for this conversation. But she was also too gracious to even think about refusing. "Oh, my, yes!" she said when I asked, with the same warmth and enthusiasm I'd seen in Lonni Sue. You couldn't imagine there was anything she'd rather do than talk to a visitor.

Maggi's father, Dwight Kennard, was a poultry scientist, she said, at the Ohio State Agricultural Experiment Station, in Wooster. Her mother, Florence, was a portrait painter who taught art at the College of Wooster during the first half of the twentieth century. "My father was very stern," Maggi said. "But my mother was very sweet and dear. She couldn't have been a more wonderful mother." Maggi wanted to be an artist as well, and Florence was the only art teacher she had, from the time she was a small child right through her first two years at Wooster.

When Maggi was seventeen, in 1935, her mother decided it was time for her to see the great works of Renaissance and Medieval art in person, not just in books. She set off for Italy and France, taking Maggi with her. "My dear grandfather paid for me," Maggi remembered. "We went to the little hill towns in Italy—Perugia and Orvieto—and saw Giotto's work and Botticelli's, and it was a thrill." This was at the time of the dictator Mussolini, and when the women got to Rome, they hired a taxi driver to find them a cheap little pensione to stay in. The dictator's palace was nearby, and Maggi remembers that the driver was very fearful about showing it to them. "I guess it wasn't really the best time to be there," she said, thinking back.

Even though Maggi had grown up in Wooster, she wasn't the most conventional of early-twentieth-century small-town Ohio girls. "In high school," she said, "I was roller-skating and climbing long before it was considered appropriate for young women. I'd try to get the cat to go up with me so I could read to it in this enormous tree we had at the house." When she was a junior in high school, she took a physics course, and, she said, "I just studied as hard as I could." She had two reasons for this. One was the material itself, which she loved. "But I was also the only girl in the class," she said, "and I had to beat the boys." She did. Just two students in each high school were allowed to compete for a statewide physics prize, and Maggi ended up being one of them. The other was a boy, of course, but she was ranked higher.

Unfortunately, the statewide test had a section on electronics. "I didn't have electronics in my background," Maggi said. "So I did very badly." She'd been toying with the idea of majoring in physics, but this setback convinced her otherwise. She went back to Plan A. Maggi studied art at Wooster for two years, at which point her mother insisted she get out of town. "If you're going to get somebody else's point of view," Florence told her, "you've got to go somewhere else." Maggi did some research and decided that the Pratt Institute, in New York, was the best option, so off she went.

Her brother, Dwight Jr., four and a half years older than Maggi, was already in New York. He'd gotten a degree in physics from Wooster, gone on to MIT for graduate school, and was now working at an engineering firm in the city. "He had a job near enough to Pratt that we had an apartment together the first year I was there," Maggie said. "That was a big help, because going to the big city after a small town was quite something." Pratt was in the Fort Greene section of Brooklyn, which was an unsafe neighborhood back in the late 1930s. Maggi and Dwight would occasionally hear gunshots. When she worked late in the studio she would run home rather than walk.

It was at Pratt that Maggi met Eddie Johnson. Eddie's Plan A hadn't involved physics or art or going to college at all. It never occurred to him, really. His father, Oscar Johnson, had emigrated from Sweden to Canada with a fourth-grade education in the early 1900s, but he was extremely well read and insatiably curious. He picked up whatever work he could—as a cowboy, then as a gravestone carver. For a while, he had a little tobacco store. Eventually he moved down to West Hartford, Connecticut, where there was a small Swedish community. He got into real estate and, at the age of fifty, finally married. His wife, Maria, was twenty-five years his junior. Eddie came along in 1919.

Henry Molaison was born in Hartford six years after Eddie Johnson, so it's conceivable that the two boys crossed paths on the streets of the city, or stood in line together for movie tickets on a Saturday afternoon. It's impossible to know if such an encounter ever actually happened or, if it did, whether they exchanged brief looks, or even had a conversation about the Lone Ranger or Flash Gordon—the younger boy destined to become the most celebrated amnesic in history, the older destined to be the father of his successor.

When Eddie was nine years old, the Great Depression destroyed Oscar's real estate business, forcing Maria to go to work as a housekeeper for a wealthy Hartford family. Eddie graduated from high school at sixteen; he was so bright that he'd skipped a couple of grades along the way. There was no thought of college, however, which was an unimaginable luxury. Eddie didn't even consider it, and no high school counselor had urged him in that direction. Instead of college-prep courses, Eddie took typing and accounting and shorthand. He had to make a living. With the economy in a free fall, however, even these practical skills didn't do him a lot of good. To help out the family he took a night job loading and unloading mailbags at the train station. During the day, he built ham-radio sets. Eddie, it turned out, had an intuitive gift for electronics. He was a natural engineer from the time he was a kid. When he was only four, in fact, he and Oscar built an electric motor together, in an odd foreshadowing of the assembly-line work Henry Molaison would take up after his surgery three decades later.

While Eddie slung mailbags and tinkered with wires and vacuum tubes and Morse code, a couple of his high school friends were determined that he could do better. They'd gone on to college, to the Pratt Institute, where Maggi was studying art. "I knew them vaguely," she recalled seventy-five years later. When the pair came back to West Hartford on Christmas break, they cornered him. "You need to go to Pratt," they said. "They've got an engineering school, and we could all room together to save money." Eddie, who still had no career prospects, didn't need much convincing. He went back to high school to take the college-prep courses he'd missed, while continuing to load mailbags at night to save up for tuition. He didn't sleep much, and he had to do his homework during class, hiding it under his desk so the teachers wouldn't notice.

But he got himself to Pratt. Maggi, two years older, was already in her final year. Those two friends of his thought they'd make a good couple, and introduced them, and, Maggi said, "We got to know each other a little bit. On Saturdays, we'd meet and walk Manhattan, walk the bridges, go to museums . . ."

It was 1941. Eddie had already joined the navy reserve, as a radio operator, mostly because he wanted to travel and couldn't afford it. He'd already been to Cuba, on a destroyer, so that looked like a good decision. Then, on December 7, the Japanese bombed Pearl Harbor. Within a couple of months, he'd been called up to active duty. Before he left, he told Maggi, "Wherever you are when I get back, I'll find you." They were on the subway when he said it, she recalled, although she could no longer say which line it was. Her hippocampus had evidently not considered that detail especially crucial, so it wasn't woven in with the rest of the memory.

They wrote frequently during the years he was away, a time when Maggi was beginning to make the transition from being primarily an art educator, which her Pratt degree had trained her for, to being a working artist. She'd already been leaning in that direction when, she said, "my dear mother sent me a tiny little clipping this big"—Maggi held her thumb and forefinger about an inch apart—"that she'd noticed in an art magazine." The following summer, said the blurb, the Spanish-American sculptor José de Creeft would be teaching a workshop at Black Mountain College, a socialist-leaning arts school founded in 1933 in North Carolina. Josef Albers, a leader of the German Bauhaus movement who had run afoul of the Nazis, had been brought in as the college's director, and Willem de Kooning, Merce Cunningham, Martha Graham, John Cage, and many other well-known artists, dancers, and musicians would end up teaching there before the college folded in 1957. Maggi's own work was primarily sculpture at the time. "I'd been hacking away at rock," she said. "I just loved stone. So I jumped at the chance to go." She didn't know a lot about Black Mountain. She did know who de Creeft was, of course—any sculptor working at the time did. She also knew about Albers, who was a major force in the art world, but she hadn't realized he was at the college. In the end, it was Albers who made that six-week residency a profound experience for her. "He changed my life," Maggi said.

Maggi remembered Black Mountain as a militantly egalitarian place. There were no status distinctions between faculty and students. One day, for example, she sat down in the dining room at a table for two. Walter Gropius, the founder of the Bauhaus school and now, having fled from Hitler, the dean of architecture at Harvard, came along and sat with her. "I had lunch with Walter Gropius!" she said, still marveling that such a thing was possible. "What an experience that was for me. It was just kind of a dream world." Albers, meanwhile, had a standing policy that any student who wanted him to review a portfolio was welcome to ask, even if he or she wasn't in one of his classes. Maggi had been out climbing mountains in Colorado earlier that summer, where she'd done some watercolors. She brought Albers photographs of those, and also of some of her sculptures. She thought she was doing pretty well. She was now teaching at Drake University. She hoped he'd be impressed.

"But oh, golly," she recalled. "He took one look at my work and said the lines of the sculpture were weak, blah, blah, blah, and then when he looked at the watercolors—awful! He scrunched my self-esteem, if I ever had any." Maggi felt totally deflated for . . . about an hour, maybe two. Then it dawned on her that maybe this guy had something to teach her. From then on, she said, "I was like a wet sponge. I took in everything he had to say, and went for it, and it was absolutely basic, basic, basic, and wonderful." Albers had strong ideas, especially about the relationships between shapes and colors. "He taught me to see in a way I never had before," she said. "What a guy, what an approach to seeing . . ." Albers changed the way Maggi thought about and taught and created art. She would go on to become not just an art educator but also a well-known sculptor and printmaker, with work in the collections of major museums and a long list of other prominent institutions. She would make art for the rest of her life.

Another instructor at Black Mountain changed her life in a more immediate and practical way. Victor D'Amico, who came to give a guest lecture at Black Mountain, was the head of arts education at MoMA. After the lecture, Maggi said, "I must have asked some good questions," because D'Amico asked if she would come back to New York to work for him. He had a program that taught art to the veterans who were now starting to return from the war. She couldn't do it right away; she had a contract to teach one more year at Drake. But she desperately wanted to get back to New York, so when the contract was up she contacted D'Amico. He remembered her perfectly well, and told her that he still had a job for her if she wanted it. She began teaching part-time at the museum and part-time at Pratt. When Eddie Johnson finally left the navy to finish his degree, in 1947, they reconnected. Maggi had been dating someone else, but she "got rid of him," and she and Eddie began walking the bridges of Manhattan again, getting reacquainted. In 1949, they were married.

By 1950, the couple had moved from New York to Princeton, New Jersey, where Eddie had been offered a job at RCA's David Sarnoff Research Center, just outside of town. RCA, the Radio Corporation of America, was the place where television was invented, and it was the original parent company of the NBC radio and TV networks. Eddie had intended to go to graduate school. Just to cover his bases, though, he went down to Sarnoff for an interview. The lab director convinced him he'd learn more on the job than he ever would in school. He never regretted the decision: he went on to hold thirty patents, run RCA's Tokyo research lab for seven years in the 1970s and 1980s, see the digital revolution happen from the inside, and play with electronics until his death, in 1989, of prostate cancer.

In the aftermath of her own illness, Lonni Sue remembered her father—they were very close—but had no memory that he had passed away. Maggi and Aline had to break the news to her. Then they had to break it again. And again. And again.

Aline remembers the "Daddy died" conversation word for word. She recounts it as though she were doing a dramatic performance for a class of third-graders, complete with exaggerated intonation, facial expressions, and hand motions.

"We were at the nursing home," Aline explained, "and Lonni Sue would ask, 'How is Daddy?,' and my mother would say, 'He died.'

"And Lonni Sue would react with shock. 'He *died*?'" Aline's eyes opened wide for emphasis. "'When did *that* happen?'

"And my mother would say, 'Eighteen and a half years ago.' That was a double shock—that it had been so long.

"'Eighteen and a half *years* ago, why didn't I *know* that?'"

And then Lonni Sue would ask again, "How's Daddy?" "He died." "He *died*? When did *that* happen?" "Eighteen and a half years ago." "And again it was a double shock," Aline said. "'Why didn't I *know* that?'"

Things could go on this way for a dozen rounds or more, and then for another dozen the next day. It's a wonder Aline and Maggi didn't lose their minds.

*Chapter 3*

# WHERE DOES MEMORY LIVE?

The destruction of Henry Molaison's memory-forming apparatus, sucked out through a tube and thrown into a hospital trash bin, came at a crucial time for neuroscience. By the middle of the twentieth century, researchers had come up with several basic questions about what memory is and how it works, but they had no good answers. They didn't know what shape memory actually takes, for example—that is, what happens in the brain when a new memory is created. They didn't know where in the brain memories are formed, or where they're stored. And they didn't know whether short-term memory, which lasts for a few minutes, is different in some fundamental way from long-term memory, which can last a lifetime. The brain damage inflicted on

Henry Molaison would ultimately do more to answer these questions than any other single event in the history of neuroscience.

Until Henry came along, for example, it was a matter of great debate whether our memories live in one specific brain location or are spread uniformly throughout the three-pound mass of spongy tissue. The theory that brain function is localized arose out of the research of Pierre-Paul Broca and Carl Wernicke, in the 1800s. Broca, who worked at the medical school at the University of Paris, was an early supporter of Darwin ("I would rather be a transformed ape than a degenerate son of Adam," he wrote in his memoir) and one of the first surgeons to experiment with hypnosis as an anesthetic. Broca's insight that specific parts of the brain have specific jobs came from the case of a shoemaker named Leborgne, who became unable to speak, or even to write, after he had a stroke, although he could still understand spoken language perfectly.

Broca got to Leborgne just in time. The stroke had happened in 1861, two decades before Broca examined him; a week after the exam, Leborgne died. The autopsy showed that the stroke had damaged part of Leborgne's left frontal lobe—the part of the brain we now know is responsible for judgment, language, problem solving, and impulse control.

Broca went on to autopsy eight other patients who had language deficits like Leborgne's and found that all of them had similar damage, in a region still known to neuroscientists as *Broca's area*. Then, in 1879, the German physician Carl Wernicke, working at the Berlin Charité teaching hospital, found another area in the back of the cortex where damage affected a patient's ability to understand language. People with injuries to what's now known as *Wernicke's area* can speak just fine—but what they say is nothing more than senseless babbling.

These two discoveries proved that at least some mental functions are located in specific parts of the brain. Another insight came from the case of Phineas Gage, Henry Molaison's rival for the title of most famous neuroscience patient in history. Gage was a railway construction worker whose frontal lobes were damaged when an explosion sent a steel rod, nearly four feet long and an inch and a quarter in diameter, shooting up like a javelin through his left cheek and out the top of his skull. He somehow survived, but, according to some reports, the accident changed him from a responsible citizen to a reckless, immoral man who inevitably yielded to his baser impulses. "The man was gross, profane, coarse, and vulgar to such a degree that his society was intolerable to decent people," read an 1851 article in the *American Phrenological Journal*. Historians have argued that the reports of these changes were probably exaggerated, but thanks to their vivid nature, Gage is just as much a fixture in psychology textbooks as is H.M.

But when it came to memory, experiments in the late 1800s and early 1900s were equivocal about whether this function was localized. In the 1920s, for example, the Harvard psychologist Karl Lashley concluded, based on studies of rats trained to navigate a maze, that memory was processed all over the brain. Lashley would start by letting the animals learn the maze; with each practice session they remembered the layout better and better, making fewer mistakes and getting from one end to the other faster every time.

Then Lashley would surgically remove different parts of their cerebral cortices. After twenty days, when he ran the rats back through the maze, it turned out that the rats remembered poorly, but it didn't really matter what part of the brain he had damaged. All that mattered was how much brain tissue he removed. "It is certain," Lashley wrote, "that the maze habit, when formed, is not localized in any single area of the cerebrum, and that its performance is somehow conditioned by the quantity of the tissue which is intact." Lashley called this idea the *theory of mass action,* which sounds more like it should have to do with physics than with neuroscience. In 1950, practically on the eve of Henry Molaison's surgery, Lashley would also write, somewhat helplessly, "I sometimes feel, in reviewing the evidence on the localization of the memory trace, that the necessary conclusion is that learning just is not possible. It is difficult to conceive of a mechanism which can satisfy the conditions set for it. Nevertheless, in spite of such evidence against it, learning does sometimes occur."

Other experiments, however, suggested that memory *is* localized. The brain surgeon Wilder Penfield, who would one day send his graduate student Brenda Milner to try and understand precisely how badly surgery had damaged H.M.'s memory, came to this conclusion. Penfield was mostly focused on using neurosurgery to cure epilepsy, but in doing the cutting he wanted to cause as little extra brain damage as possible. Since the brain itself can't sense pain, these operations were done only with local anesthetics. The patients were awake the whole time. That way, the surgeon could deliver a tiny jolt of electricity to stimulate specific points in the brain near the tissue to be removed, to see whether they served some crucial function, and should thus be avoided with special care. Penfield discovered that these electric jolts sometimes triggered vivid memories—one patient reported a memory of a dog chasing a cat, for example, while another heard a familiar voice, saying words the patient couldn't quite make out. But this happened only when the electricity was applied to the temporal lobes. These, Penfield concluded, must be the seat of memory.

In a sense, all of these scientists turned out to be right. Memory is localized in the temporal lobes, but it's also located everywhere in the brain. This seems like a paradox, but it's not.

The kind of explicit, autobiographical recollections most of us refer to when we use the word "memory" are made up of sounds, smells, sights, sensations, and emotions. When Aline Johnson approached me on the street to tell me what had happened to her sister, my mind called up impressions I formed in the late 1960s: the sight of Aline, the sound of the music teacher's voice calling her name, the sensation of dread that would grip my intestines when my classmate launched into that perfectly executed drum roll signaling that I was about to fail again. All of those impressions had originally entered my brain through different routes—my visual system, my auditory system, my somatosensory system (which tracks bodily sensations), my emotion-processing apparatus. The physical traces of each impression remained lodged in whatever part of the brain had processed that aspect of the experience. Memories of sights are stored in the visual-processing system, memories of sounds in the auditory system, and so on. Lashley's rats could keep navigating their maze after he cut out chunks of their brains because they'd been finding their way using multiple cues, not just vision or smell alone.

But the damage inflicted on H.M.'s brain showed that the processing of memory was also localized. The elements of memory might be stashed in many different parts of the brain, but the removal of one specific part was nevertheless devastating to Henry's ability to tie them together and access them. He could hold a new thought in his head, but only for a very short time. After that, it was gone. This was clear enough to his parents and to the hospital staff who dealt with him after the operation. It wasn't clear precisely how short his retention time was, however, until Henry and his mother took the train to the Montreal Neurological Institute, in 1962, nine years after the surgery, for a more intensive follow-up to the testing Brenda Milner had done during her visit to Hartford in 1955. By now, Milner had begun her own research program at the Neuro, as it was known. For a week, she and her colleagues put Henry through experiments, while his mother sat in the waiting room for hours at a time. (According to Corkin, who had become Milner's graduate student the year before, Elizabeth Molaison didn't venture out to explore Montreal because she was "terrified of the big city where people spoke French, a language she did not understand.")

Henry underwent dozens of tests during that week, including one that explored the extent of his short-term memory. The idea that such a thing exists, and that it's somehow different from long-term memory, dates to the late nineteenth century. In 1890, the Harvard psychologist William James proposed in his book *The Principles of Psychology* that memory comes in two forms, which he called *pri-*

*mary* and *secondary*. The first was his term for short-term memory—the awareness of what's going on around us at the moment, and of what has been happening just before. The second referred to long-term memory—what James described as "the knowledge of an event, or fact of which meantime we have not been thinking, with the additional consciousness that we have thought or experienced it before."

James had had no experimental evidence to back up his proposal, and he never tried to pinpoint exactly where short-term memory left off and long-term memory began, but many scientists who came afterward found the distinction broadly convincing. Now Henry's case appeared to bear it out. The fact that he could carry on a conversation obviously meant that he had at least some short-term memory. He could respond appropriately to questions, for example. "If I asked him, 'Henry, did you have a seizure today? What did I just say to you?'" Corkin told me, "he would reply, 'You asked me if I had a seizure today.'" But if she asked him what she'd said an hour ago, he had no clue.

A person with normal memory function might not remember either: the information would have vanished from his or her short-term memory to make room for something new. But it wouldn't be gone entirely. The information would have transitioned into long-term memory, from which it could be retrieved, given the right cues. Most of us have had the experience of walking into a room purposefully but forgetting what the purpose was. When you go back to your starting point, though, simply being in the original location will often be enough to remind you of what your original mission was. You might not remember what I said to you an hour ago, but if I remind you of the topic, it will probably come back.

Even if you gave Henry a dozen hints, however, or walked him into a dozen rooms, he still wouldn't be able to recall. His brain wasn't moving his short-term, immediate memories into long-term storage. With Henry, scientists now had the opportunity to figure out how long it took for that transition to happen.

To do that, Corkin says, a grad student named Lilli Prisko exposed him to a set of stimuli—flashes of light, or patches of color, or geometrical shapes, or musical tones, or rhythmic clicks. Then, after a delay, she'd repeat the stimuli, and Henry would have to say whether the second version was the same as the first or different. In the case of the light flashes, they might differ in timing—three per second the first time but faster the second time, or maybe slower. The color sample might initially be bright magenta, then crimson. Or maybe it would be magenta again. Henry didn't have to characterize the difference; he simply had to say whether there was one.

Unsurprisingly, he did pretty well when the second exposure came immediately after the first. He did about the same when it came fifteen seconds later. He did reasonably well after thirty seconds. But when the delay between the first and second exposure was as long as sixty seconds, Henry's performance tanked. His effective memory span, it turned out, was somewhere between thirty and forty seconds. Yet he could still remember facts and general information he'd acquired decades earlier. Thanks to Henry, William James's dichotomy between short-term and long-term memory had been proven correct.

Henry was also instrumental in demonstrating another sort of split that James had proposed in *The Principles of Psychology,* between the memory of facts and the memory of skills. James called the former *memory* and the latter *habit,* and they're clearly different. The example I always use in my own life is tying a bow tie. The Latin department at Princeton High School had an annual tea, where the students who had done especially well at translating Cicero (or whatever) received awards. I took Latin, so I was forced to go. Even in 1970, the idea of teenagers attending a "tea" was hopelessly archaic for a public high school. Only the Latin teachers would put on such a ridiculous thing.

The day before the tea, I happened to go to a rummage sale at the local firehouse, and spotted a bow tie. Not a clip-on, but the real thing. At the time, bow ties were considered the essence of goofiness, and I knew instantly that I had to wear this one to the Latin tea, as a commentary on the dorkiness of the event. The problem was that I had no idea how to tie it. I instinctively knew who would, though: my English teacher, William Cook, a brilliant, charismatic, and elegant man who would go on to become a professor at Dartmouth. I have a vivid memory of finding him in his classroom after school, asking if he could show me what to do. Of course he could, and I walked out of that room with a new skill. What I don't remember is how he explained it. Forty-five years later, I can still tie a bow tie without hesitation. I can't teach someone else, however, because I can't translate that skill into words. At about the same time, however, my friend Jim taught me how to drive a car with a manual transmission. That, too, is something I can now do instinctively. But since I remember in this case exactly what my instructor told me to do, I can teach anyone else as well.

When she first tested Henry Molaison during her visit to Hartford, in 1955, Brenda Milner showed that these two forms of memory, which had been described variously as memory versus habit, explicit versus implicit memory, "knowing that" versus "knowing how," and declarative versus procedural memory, were processed by the brain in different ways. The experiment that showed this was so elegant and so convincing that it appears in pretty much every introductory psychology and neuroscience textbook. Milner showed Henry a piece of paper with a five-pointed star printed on it, and asked him to trace the image, on the same piece of paper, staying just inside the star's outline. The catch was that he couldn't see his hand or the star directly, but only as reflected in a mirror.

Like anyone else, he found it nearly impossible at first. When he moved the pencil to the right, its image moved left, and vice versa. There was a 180-degree disconnect between the information coming in through his eyes and the information coming in through his hand. It was bewildering. When Milner repeated the experiment a day later, Henry had no memory of ever having done it before—but he traced the star more easily than he had on the first day. When she repeated it on a third day, he still had no memory, but he'd gotten even better. As Milner later told Suzanne Corkin, Henry was surprised. "Well, this is strange," Milner recalled him saying. "I thought that this would be difficult, but it seems as though I've done it quite well."

Seven years later, in Montreal, Corkin herself put Henry through a test that showed the same thing, but in a more rigorous, quantitative way. She had him trace a path through a maze eighty times in a row. He never learned the maze itself. He made just as many mistakes before finding the correct route on the eightieth trial as he had on the first. But the time it took him to get from start to finish steadily improved. The layout of the maze hadn't become part of his explicit memory, but the knowledge of how to navigate a maze was stored in his procedural memory. Since the hippocampi and some surrounding tissues in Henry's medial temporal lobes were no longer part of his brain, they must be necessary for the former but not the latter.

Milner and Corkin and dozens of other scientists went on to give Henry hundreds of tests over the next forty years (and Henry was almost always extraordinarily patient and cooperative). They would gradually begin to understand that the categories of memory Brenda Milner and her students first identified in the 1950s and 1960s were highly oversimplified. The boundaries between procedural memory and declarative memory—between things we "remember" unconsciously and the things we remember consciously—are fuzzier than those early tests on Henry and on other amnesia patients first seemed to show. Even within conscious, or declarative, memory it's hard to draw a sharp line between episodic memories of general facts and semantic memories of specific incidents.

Henry also helped neuroscientists to establish that still other forms of memory exist that don't fit neatly into any of these categories. For example, he could retain new information through an unconscious process called *perceptual learning.* Corkin and Milner gave him something called the Gollin incomplete-figures test, which starts off with drawings of twenty common objects, rendered in such a sketchy way that you'd never be able to tell what they were. Subjects have to guess anyway. Then they get the same objects again, but this time the drawings are a little more complete. They guess again. They get versions of the objects that are more and more complete until they can identify all twenty. According to Corkin, Henry did a little better at this than normal control subjects. His perception was intact, but that's not what this test was about.

An hour or so after going through this exercise, she and Milner gave Henry the test once again. He had no conscious memory of ever having seen it before, of course, but he figured out what the images were more quickly this time. Even more surprising, he retook the test thirteen years later, and even after such a long delay he did better than he had the first time. The memory of having done this before was in his head somewhere, even though he didn't know it. This isn't procedural memory. It's not a memory of how to use muscles in a skilled way, as the maze and drawing tests were. But it isn't exactly declarative, either. (Corkin also points out that while Henry got better at identifying the images, he didn't improve as much as the normal control subjects did. This was probably because they had a conscious memory of what pictures they had seen during the first round. They remembered what the available choices were.)

Henry's unconscious memory was also intact for something known as *repetition priming.* The scientists showed him a list of words and asked him to identify something about them—whether they contain the letter "A" or not is the example Corkin cites in *Permanent Present Tense,* her memoir about working with H.M. The thing he was asked to identify wasn't directly relevant to the test; it was only so that he'd pay close attention to the words. Then they showed him the first three letters of the same words, along with the first three letters of words he hadn't seen, and asked him to guess at what the full word might be. He guessed right more often when the three letters came from the words he'd recently seen, showing that he remembered more than he consciously realized.

Yet another form of memory Henry retained was the sense that someone or something is familiar, as in "I know I've seen this person before, but I have no idea where or when." This isn't procedural either, yet it's not entirely declarative. Yes, you can declare that something is familiar, but you can't necessarily explain why. Despite the catastrophic loss of his hippocampus, Henry retained this form of memory. Neal Cohen, a neuroscientist at the University of Illinois in Urbana-Champaign, experienced this in a concrete way. In the 1980s, Cohen worked with Henry for three years as a postdoctoral fellow in Corkin's lab at MIT. "That was what drew me there—the chance to work with Henry," he said.

During that time, Cohen would frequently be the one who ferried Henry from Hartford, Connecticut, to Cambridge, Massachusetts, and back for testing. Cohen was also deeply involved in the testing itself. "I would see Henry for hours every day for two weeks at a time," Cohen said. He'd introduce himself to Henry many times during that period, since even after two weeks Henry didn't know who he was. But if Cohen brought in a student or a postdoc Henry hadn't been interacting with and asked, "Do you know one of us?," Henry would pick Cohen every time. Cohen had become familiar. Corkin had the same experience. He was also familiar with her, but again, he didn't know exactly why that should be. If you pressed him, Cohen said, Henry would invariably say, "We went to high school together." That was the case, Cohen said, "whether it was me, or Sue, or a first-year graduate student. Didn't matter what the age of the person was, if you were familiar, he'd say he knew you from high school."

It was as though Henry's mind simply couldn't rest without resolving the mystery. It was desperate to make connections. He knew you from somewhere; he remembered going to high school in a general, semantic way; he knew there were lots and lots of people he should be familiar with from that period in his life. Therefore, he must know you from high school. Most of us have that same impulse to solve mysteries of familiarity. "I just *know* I've seen him somewhere before," we think, and keep jiggling our memory, trying to figure it out. Henry didn't understand that he had profound amnesia, since he had literally forgotten what it was like to have a normal memory, but his mind still had that impulse. He knew that high school was a plausible answer, so his mind convinced him that it was the correct answer. In the language of neuroscience, he was *confabulating*—making stuff up that seemed plausible, then convincing himself that it was true.

In one sense, it turned out to *be* true. While Cohen was working in Corkin's lab, he said, he, Corkin, and another neuroscientist, named John Gabrieli, decided to take Henry to his thirty-fifth high school reunion at East Hartford High School. "We walked around with him," Cohen said, "to see who he actually remembered, and what that experience was like for him." Unfortunately, the scientists realized only belatedly that Henry had gone to two different high schools, and they picked the graduating class he was less familiar with. "It really wasn't the people that he had traveled through school together with."

Nonetheless, Cohen said, Henry did sort of know one or two people who had been in homeroom together with him, and some of them knew him. "It was a powerful experience for all concerned," Cohen said. There was also a photographer at the reunion. Toward the end of the evening, he shepherded everyone together for a class picture. The scientists quickly got out of the way, of course. "We went into some corner," Cohen said. As a result, they weren't in the final group shot, but they do appear in some of the reunion photos. "So it turns out Henry was right all along," Cohen said. "Apparently, we did go to high school together."

*Chapter 4*

# PRINCETON

When they first moved to Princeton, the Johnsons didn't own a car. Maggi took the train into New York, to continue teaching at MoMA, while Eddie would ride his bicycle to work across U.S. Route 1—now a major, traffic-choked artery packed with malls and office parks and pharmaceutical labs, but back then a sleepy rural four-lane highway lined mostly with woods and sod farms. The town of Princeton was perfectly described by Albert Einstein, who'd come to live there in 1933, and who was still shuffling around town, shock-haired and sockless, when the Johnsons arrived. It was, he wrote, "a wonderful little spot, a quaint and ceremonious village . . ." The rest of the quotation, which comes from a letter Einstein wrote to Queen Elizabeth of Belgium, reads "... of puny demigods on stilts." That last phrase was addressing the town's old-money aristocrats. Princeton is somewhat less quaint today. It's overrun with tourists who are disgorged by the busload on summer mornings to gawk, among other things, at the house where the great physicist lived. The family-owned grocery, hardware, and clothing stores Einstein used to patronize have mostly been replaced with expensive boutiques and high-end chain stores like Brooks Brothers. The aristocrats are fewer as well, but sharp-eyed natives can still spot them.

In 1951, Eddie and Maggi built a house at the northeastern edge of Princeton, which at the time was still mostly a mix of woods and open fields. Lonni Sue had been born about a year earlier. She'd been given a name that seemed very out of place in an educated, cultured town filled with Einstein's puny demigods. The way Maggi told it, she wanted to give the baby a unique first name, because "Johnson" was so common. She played around with several, and finally settled on "Onni," which she had invented. She liked the sound of it. Eddie would do almost anything for his wife, but he wasn't willing to go quite that far. The name was just too weird. So she stuck an "L" at the front, making it "Lonni." But the newborn girl looked more like a Sue—so they added that as an afterthought. Aline was born four years later, and named after Aline Saarinen, a *New York Times* art critic Maggi admired, who was married to the architect Eero Saarinen. The family called her "Lini" for short.

Lonni Sue no longer remembers her childhood, but in 2004, three years before the encephalitis struck, she wrote a brief introduction to a catalog for one of Maggi's exhibitions that makes at least part of it clear:

> My favorite vision of my mother from childhood was of her in her apron, standing between the kitchen and studio, on the threshold, with her hands on her hips, between being a mother and being an artist. She liked to make prints more than cook, and her apron had more inky fingerprints than spaghetti stains.

It made all of the difference. It was the best mix of mothering of all. We were well fed, with both food and inspiration.

The age gap between the girls was wide enough that they were never especially close growing up. By the time Lini was a toddler, Lonni Sue had gone off to kindergarten. They overlapped for one year during elementary school, but aside from that they went off in different directions on weekday mornings, and never attended the same school again. They had entirely different sets of friends, naturally. There were a lot of kids in the neighborhood who roamed around in packs, going from one house to another, playing in one another's backyards or out on the quiet streets. The Johnson girls weren't really part of it, however. They would spend most of their free time in their rooms, reading, although Lonni Sue would sometimes play with a girl named Emily Speagle, who lived next door. "If we hadn't been neighbors," Emily told me, "I don't know that we would have been such friends." The Johnsons had a big maple tree in the backyard with a patch of sand and dirt surrounding it where grass wouldn't grow. "We'd just kind of creatively play," she said. "So we were buddies in that sort of sense."

The one thing the sisters had in common was music. Eddie Johnson had fallen hopelessly in love with classical music sometime during college. He had to have it playing during every waking moment, at least at home. Years before sleep timers were available on clock radios, Eddie rigged up his own electronic contraption that would shut off the stereo after he fell asleep and turn it on again in the morning to wake him up. "I never knew whether he played music at the labs or the workplaces," Maggi said, "but he always had to have it at home, and so the girls heard music before they were born, if that makes any difference."

It very well might have. Both girls began piano lessons in first or second grade, and as soon as instrumental music was offered in Littlebrook School (as in all of the Princeton public schools of the time, this happened in fifth grade), Lonni Sue took up the violin, then eventually switched to viola. When Lini reached fifth grade, she chose the cello. Their music teacher quickly realized in both cases that the Johnson girls were unusually talented. He insisted that they take private lessons. Eddie and Maggi wisely agreed. Both girls turned out to be extraordinary musicians. As they got older, the Johnson home turned into a sort of arts and culture salon. Curt Carlson, a graduate student in electrical engineering at Rutgers University, about ten miles north of Princeton, first heard about Lini and Lonni Sue through a musician friend named Jim Scott. "I know this wonderful family in Princeton," he remembers Scott saying. "They've got a daughter who plays the cello beautifully, and another who's a terrific violist. Let's go play music with them."

Curt Carlson was a very talented amateur violinist himself, so they went and played Mozart piano quartets, and he quickly became good friends with Eddie and Maggi. Carlson would ultimately go to work at Sarnoff as well. Carlson's future wife, Dudley, heard Lini's and Lonni Sue's names long before she met them. "He was clearly just smitten with the whole family," she says. After Curt and Dudley married and moved down to Princeton themselves, they would frequently go over to the Johnsons for dinner followed by an evening of chamber music. When Eddie Johnson died, this casual quartet—Curt Carlson, Jim Scott, and the two girls, now all grown up—played for his memorial service. The girls thought they couldn't handle it, but Curt knew they could. "How can we *not* play?," he asked, so they did. Maggi said afterward that they'd done just beautifully.

Aline still remembers the conversations that bubbled during those string-quartet dinners, and the equally effervescent conversations at dinners where it was just the four Johnsons. In both cases, the topics would range from art to science to history to politics to anything else someone had on his or her mind. "It was a sort of family philosophy that you must be interested in everything," she says. This began even before the girls were born. "Eddie didn't know a thing about art when we met," Maggi said. "He didn't have a clue. He was so eager to learn things that as time went on he got to be my best critic. And I loved to hear about what he was working on, even though I didn't understand half of it."

Dudley Carlson, who would become a beloved children's librarian at the Princeton Public Library, experienced the dinner talk with a sense of moderate awe. It was always a sort of many-player Ping-Pong match, she says, with somebody putting up an idea and everybody taking shots at it and pulling it apart, and looking at it, very much like the give-and-take that comes with the playing of music. "What's this really about? What's going on here? But what do you think? What do *I* think?" she said, trying to capture it.

The Johnsons themselves were a quirky quartet, but Lonni Sue was perhaps the quirkiest among them. "It's a word I've always associated with her," Carlson says. She had a kind of elfish laughter, and a sense of playfulness that came out, not just in the way she approached music, but also in the way she thought about language. She was endlessly fascinated with the meanings of words. "She's always been a wordplay person, a punster," Carlson says. Lonni Sue's own memories of those dinners and chamber music sessions vanished almost completely with her illness, destroyed by the virus that ravaged her brain. But the wordplay and puns somehow managed to survive. During our first meeting, not long after Aline had approached me on the street, Lonni Sue looked at me and said, "Your face is so nice! 'Face' has the word 'ace' in it. It's wonderful for me to see all the words that have 'ace' in them. Have you ever tried that?" I never had, I admitted. She asked my name, and I told her it was Mike. Not missing a beat, she asked, "Do you use a microphone?" That's how a conversation with Lonni Sue goes these days—although it evidently leaned in that direction before she got encephalitis. Her wordplay is just as sophisticated as it was before she got encephalitis, but since it consists almost entirely of puns, that's not necessarily saying much. It's quick and clever, but not especially deep.

Carlson also remembers another Johnson family trait. "Right

from the start," she said, "it was clear that the Johnsons were the most positive, upbeat, cheerful people I'd ever met." Maggi, in particular, seemed incapable of being negative about anything at all. She herself attributed this to a time when she was a very little girl, tagging along with her brother and his friends. She began whining that she wanted to go home. Dwight took her aside and told her that "nobody likes a complainer." She evidently never forgot that lesson. Just a few years ago, having never stayed in a hospital for anything other than childbirth, Maggi was told she needed a hip replacement. She was naturally unhappy about the prospect, and Aline, who had seen her mother find the good in every situation for at least fifty years, was convinced Maggi couldn't pull it off this time. Surely, she would see her mother's spirits sink for once. Aline was wrong. "Well," Maggi said brightly, "maybe it's a good opportunity to see what being in the hospital is like."

Despite everything she's gone through and everything she's lost, Lonni Sue's mood is almost always equally upbeat. Like Henry Molaison, she's largely unaware of how much she's lost. Still, you might reasonably expect that the catastrophic damage to her would be so disorienting that she'd be constantly depressed, or anxious, or even terrified. What's happening?, she might wonder. What is this place? How did I get here? Who are these people who keep talking to me? They act like they know me, but I've never seen them before. Why can't I remember? That's true for some amnesia victims, but not Lonni Sue. When she greets you, it's as though your appearance is the nicest surprise she's had in months, and if you exit the room and return, it's the nicest surprise all over again. She laughs easily and often. She uses the words "inspiring" and "beautiful" and "wonderful" frequently. Every so often she'll break spontaneously into song. She's having a blast, and she wants you to join her.

By the time the Carlsons were getting to know the family, Lonni Sue had begun making art as well as music. She'd been drawing long before that, as most children do, but while she was good at it, she wasn't especially creative. "She drew horses, horses, horses," Maggi remembered. She also read about horses, and Eddie built a series of hurdles in the backyard so she could jump over them, pretending she was a horse—or pretending that she was riding one. Nobody seems to remember which it was, exactly, but the practice eventually got her a spot as a hurdler on the high school track team. Eddie and Maggi let Lonni Sue take riding lessons at a local stable, where she would cheerfully dig out manure from the stalls, groom the horses, and generally hang around.

The real horses were fine, although her parents did worry a bit that Lonni Sue might fall, break her arm, and be unable to play the viola. The horse drawings, however, began to drive Maggi slightly nuts. Lonni Sue's artistic talent was already becoming clear, and it distressed Maggi to see her daughter focusing exclusively on such a cliché. She knew that trying to push Lonni Sue would almost certainly backfire. Pulling, however, was another matter. The agony finally ended when Maggi decided to teach a summer art class for high school students for the Princeton Art Association—specifically so that Lonni Sue could take it. It was a version of the basic design class she'd been teaching since that extraordinary summer at Black Mountain College a quarter of a century earlier. Josef Albers had given Maggi an entirely new way of looking at the world, and it transformed her art. It's not just a bunch of lines and colors and textures; it's a medium of communication, a language with its own vocabulary and grammar and syntax. Now Maggi had the chance to pass this visual language on to her daughter. "I gave them visual problems to solve that weren't horse problems," Maggie said, "and Lonni Sue gobbled it all up. She bloomed, I have to say."

Lonni Sue went on to study humorous illustration at the School of Visual Arts, in New York, and it was at this point that her characteristic style began to emerge. It was very different from her mother's, which was mostly abstract. You might have a sense of what Maggi's art was supposed to represent, but you could rarely be certain—and often it didn't represent anything specific in the first place. Lonni Sue, by contrast, favored quirky pictures of people, of buildings, of animals. A chair would be as tall as the ceiling; fingers or fingernails on people would be elongated; objects or characters would be seen from several perspectives at once, Picasso-like but whimsical at the same time. Cats made frequent guest appearances. So did horses, which she never abandoned. So did tiny people. The sun or the moon might have a face. None of it was at all realistic, but at least there was no question about what you were looking at.

Lonni Sue's understanding of shapes and colors and their relationships to each other came directly from her mother, but the humor and visual puns that almost always populated Lonni Sue's work came from Eddie. He was the one, Maggi insisted, who had the sense of humor. An early piece of Lonni Sue's that Dudley Carlson especially loves is titled *Read Letter on the Red Letter Table.* It illustrates Lonni Sue's sensibility perfectly. It has a mostly red background, with blue above and blue below, and the shadows are all blue, but the red and the blue bleed into each other. At the center is a coffee cup and saucer, and also a letter (which has been read, clearly), unfolded on a tabletop whose existence is only hinted at. "But the color is what makes you look three times at it," Carlson says, "and the starkness of the white letter centered in a sea of color."

That piece came early in Lonni Sue's professional career. Her sophistication as an artist had evolved rapidly after the class with Maggi, so perhaps it shouldn't have been quite such a surprise when she declared her intention to major in art during that long drive to Ohio in the fall of 1968. She spent a year at the College of Wooster, then transferred to the University of Michigan, where she got a degree in printmaking and photography. She also played her viola in the university's Gilbert & Sullivan orchestra. It was a perfect fit: the operettas had the same sort of complexity and punnery and creative playfulness that characterized her conversations and visual art.

Maggi wasn't terribly impressed with the art education her daughter received at Michigan, though. It certainly didn't seem to her to be on a level with what she herself had gotten from Josef Albers and her other mentors in just a month and a half at Black

Mountain. But by now, Lonni Sue was striking out in her own idiosyncratic direction. "I think they pretty much let her be her own creative self," Maggi said. Given how her career would unfold over the next few years, it's hard to imagine how they could have stopped her.

After she finished up at Michigan, Lonni Sue was determined that she would become a working artist like her mother. She had no job and no income, so she moved back to her parents' house to try to establish herself. Eddie obligingly built her a darkroom in the basement, but he wasn't entirely happy with the situation. Despite her own success, Maggi didn't support herself purely as an artist. She made most of her living as an art educator. She had a day job. Maybe Lonni Sue could get her own day job as an art director at a magazine, Eddie thought, designing pages and arranging the visuals that decorated articles. Lonni Sue didn't like that plan at all. She wanted to make art, not shuffle other people's art around.

But she didn't want to live at home forever, either. She needed to earn some money. So after a year of living with Maggi and Eddie, she took a job teaching art at the Stuart Country Day School in Princeton. It was good to have money coming in, and besides, once she was out of the house it was a lot easier to begin networking with other artists. One of the first people she met was Bob Denby, who taught photography at Princeton Day School, just around the corner from Stuart. At the time, Denby was a trim man with short, curly hair. Nearly half a century later, he still is. He looks a little too preppy to be a serious artist, but that's deceptive. When he and Lonni Sue first met, he had already completed a documentary about Shirley Chisholm, the first African-American woman ever elected to Congress, and the first to run for president. (Chisholm ran for the Democratic nomination in 1972. She didn't get very far.)

The first time Denby met Lonni Sue, he was struck at once by what he calls her "otherworldly soulfulness," a sort of childlike wonder that was perfectly consistent with her artistic sensibility. She had a way of embracing the joy in life, he said, acknowledging immediately that this sounds utterly trite. "But that was Lonni Sue," he insisted. She was also, Denby said, the first person he'd ever met who was a truly active listener. "You'd be talking to her and she'd just look at you with this penetrating expression, stone still and silent, absorbing it. There are not many like that in this world," he said, "and we became friends."

They also became collaborators. Teaching was Denby's day job as well, but on the side he was something of an art impresario. He won a commission to do a huge mural for a local bank, and hired Lonni Sue to execute it (the mural is still there, but the tellers who stand in front of it have no idea of its history). She also began collaborating with a local merchant named Robert Landau, whose parents had founded a woolen-goods store on Princeton's main street in 1955. It's still in business, one of the few exceptions to the boutique-ization of the town.

Sometime in the early 1970s, Lonni Sue had a show in a local art gallery a few doors up the street from Landau's. (The location is now a Starbucks.) Robert Landau didn't know much or care much about art at the time. He'd certainly never bought any, and it had never occurred to him that he ever would. But he was moved to buy one of her pieces. It was titled *Dollar Factory.* Lini describes it on the Web site she still maintains for her sister. It is, she writes, "a fanciful concept of how dollar bills are produced assembly line style: Birds in a cage go down a conveyor belt to go onto an eagle stamp, workers are snipping off the tops of pyramids, and a masked figure is chopping off presidential heads."

When he found out that the artist whose work he was so drawn to was local, he contacted her immediately about drawing some covers for the Landau's catalog. Ultimately, she would do nine of them. He also had her design some advertisements. People would often ask him what his target demographic was, and he would always say it was the kind of people who read *The New Yorker*—not wealthy, necessarily, but people with wide-ranging and eclectic interests. It occurred to him that if this was true, he should be advertising in the magazine. It was also clear that Lonni Sue's quirky style would be perfect for that same audience. In the end she produced a dozen or so ads for Robert Landau, with the understanding that he would get to keep the original drawings. They still decorate the house he shares with his wife, Barbara. Lonni Sue also did the birth announcement for the Landaus' son, Matthew, who was born in 1982.

The biggest project Landau and Lonni Sue collaborated on was something called the Princeton Poster, which he dreamed up as a promotional gimmick for himself and fellow merchants. The idea was for Lonni Sue to draw a brightly colored, fanciful 3-D map of Princeton that would show town landmarks, including churches, schools, and Princeton University. It would also highlight local businesses, but only as long as they paid for the privilege. And it would include plenty of visual and verbal puns, many of which would be apparent only to people who knew the town well. You could think of it as one big inside joke. Many Princetonians did.

In the end, seventy or so businesses signed up, which Landau regarded as both an extraordinary success and a potential nightmare—an impossible project, he called it. Each of the seventy business owners had strong ideas about the image he or she wanted to project. Lonni Sue had to make them all happy while still maintaining an overall consistency in the design. She'd developed a knack by this time for working with clients: for the Landaus' birth announcement, for example, she'd come up with ten different concepts they could choose from. In this case, she had seventy times as many opinions to deal with. Yet to Landau's frank astonishment, she made it all work. No one ever did any research to figure out whether the poster generated extra sales for Princeton merchants, but it was wildly popular with the locals. Thousands were printed, and there wasn't a single one left over. For years, you'd invariably see the Princeton Poster tacked to the wall when you went into someone's house in town, and even today, you just have to mention it, and everyone knows exactly what you're referring to. "She made everyone happy," Landau says. "Part of the reason is that she was happy and joyful herself. It was infectious."

He's seen her several times since the encephalitis struck. She has no idea who he is, which is understandably distressing. Aside from that, he says, "When I talk to her now she seems as joyful and happy as she was thirty years ago."

*Chapter 5*

# HOW CELLS REMEMBER

It's impossible to overstate how important Henry Molaison's surgery was to neuroscience. Within just two years after William Scoville destroyed his hippocampus, researchers had at least rudimentary answers to questions they'd been asking for more than half a century. Henry couldn't form new conscious memories, so his hippocampi, along with surrounding tissues in the medial temporal lobes, which had been destroyed, must be crucial to that process. He could call up some conscious memories from his past—mostly general facts, rather than specific episodes—so they must be stored elsewhere in the brain. He could hold new thoughts in his head for forty seconds or so, so short-term memory must be different from long-term memory. He could learn new skills, suggesting that "knowing how," as the British neuro-philosopher Gilbert Ryle described it, involved a different memory system than "knowing that."

Henry's ravaged brain helped neuroscientists understand the categories of memory, but it didn't help them to address other longstanding questions: What form does memory take? How does the brain change when a memory is created? What is the physical substance of memory—the hypothetical entity neuroscientists in the early twentieth century referred to as the *engram* or the *memory trace*? The seminal paper Miller and Scoville wrote for the *Journal of Neurology, Neurosurgery, and Psychiatry,* in 1957, which introduced Henry's remarkable case to the neuroscientific world, didn't address these questions at all. But it nevertheless inspired a young scientist named Eric Kandel to look for the answers. Kandel had fled occupied Austria in 1939 with his family, at the age of nine. He had lived through the violence of Kristallnacht, the night in 1938 when Nazis and their sympathizers attacked synagogues and Jewish-owned businesses, and assaulted Jews themselves. It was clear that Jewish families had to get out of the country if they could. His father, Hermann Kandel, was a prosperous businessman, so they had the resources to escape what would very quickly escalate into the Holocaust.

They wound up in Brooklyn. Eric would eventually go on to college at Harvard. He was fascinated by the mind, and at first he wanted to become a psychoanalyst, like his Viennese Jewish hero, Sigmund Freud. To do that, he had to go to medical school, so after graduation he enrolled at NYU. Freud's ideas about the brain—specifically, that our lives are ruled by two competing mental entities, the id and the ego—weren't grounded in any actual biology. Freud deduced the existence of the id and the ego by pondering his own thoughts, impulses, and emotions, and by listening to his patients ponder theirs. By the time he was in his last year at NYU, Kandel had decided he wouldn't be a psychotherapist after all. He didn't want to understand the dynamics of the id and the ego; he wanted to find out where in the brain they lived. He wanted to know the mechanics of how they worked. He explained this to one of his professors, Harry Grundfest. "Grundfest listened patiently as I told him of my rather grandiose ideas," Kandel writes in his autobiography, *In Search of Memory,* going on to say that

> another biologist might well have dismissed me, wondering what to do with this naïve and misguided medical student. But not Grundfest. He explained that my hope of understanding the biological basis of Freud's structural theory of mind was far beyond the grasp of contemporary brain science. Rather, he said, to understand mind we needed to look at the brain one cell at a time.

At first, Kandel writes, this was "demoralizing." He wanted to answer grand and sweeping questions about how the mind works. Understanding a single brain cell seemed just too pedestrian.

Kandel knew, however, even as he had this thought, that it wasn't pedestrian at all. Back in 1955, when this conversation took place, many neuroscientists were beginning to understand that all mental activity arises from the electrochemical activity of tens of billions of individual cells, called *neurons,* sending messages back and forth through an unimaginably complex network of interconnections. The intense sorrow you endure at the death of your parent or your spouse or your sister or your dearest friend, or the indescribable joy a mother experiences at holding her newborn child, is simply a set of neurons talking to one another. The terror that grips you during a slasher movie, or when you're in actual danger, is really nothing more than neurons firing off. Even your awareness that you're sad or terrified or elated—it's just neurons.

It doesn't seem that way, of course. Our mental states seem so profoundly different from other bodily sensations such as hunger, pain, or fatigue that we intuitively feel that they must arise from something more than mere biology. That's what René Descartes thought. The seventeenth-century French mathematician and philosopher concluded that the mind and the body are separate, although he acknowledged that they do communicate with each other. He thought the conversation happened in the pineal gland (in fact, the pineal gland governs our sleep-wake cycle and other body rhythms). Mental activity—especially self-awareness—seems so self-evidently divorced from the physical body that Francis Crick, the molecular biologist who codiscovered the structure of DNA, titled his 1994 popular book on the biological basis of consciousness *The Astonishing Hypothesis*—that hypothesis being that there's nothing going on inside but neurons talking to one another. For the typical lay reader, the notion that all of our mental activity amounts to no more than the electrical flickering of brain cells, going off like so many microscopic spark plugs, still seems hard to believe.

Even back in 1955, it wasn't quite so astonishing to a young Eric Kandel. More than a half century earlier, the Spanish anatomist Santiago Ramón y Cajal had been the first to describe the structure of neurons in any detail, observing them through a microscope, then making exquisite drawings (he originally thought he might become an artist). Ramón y Cajal showed that most neurons have three parts. The first is a central cell body, containing the cell's nucleus. The second is a long, narrow projection known as an *axon,* which can extend up to several feet from the cell body, reaching out into the brain or down the spinal column. The axon may eventually split into branches, and the branches split further into multiple tiny ends—the *terminals.* You can think of the axon as a tree trunk that sprouts branches, which in turn split into twigs.

The third part of a neuron is the *dendrite,* which emerges from the cell body on the side opposite the axon and splits into twiggy ends as well. The cell body sprouts as many as forty dendrites but only a single axon. Ramón y Cajal knew, based on experiments by an earlier generation of scientists, that neurons both send and respond to electrical signals. He deduced from his own investigations that these signals travel down the axon and out to the axon's terminal branches. Each terminal connects to the dendrite of another cell. When a signal comes down the axon and into the terminal, it stimulates the corresponding dendrite, triggering a signal that travels to the cell body, then to the axon, and the process repeats itself. Dendrites receive electrical input; axons generate electrical output. Each cell might be connected, terminal to dendrite, with a thousand other cells, some of them in entirely different parts of the brain. With up to one hundred billion neurons—comparable to the number of stars in the Milky Way—that means each brain cell can make one hundred *trillion* possible connections. Because the brain's wiring diagram is so vastly more complex than the most powerful computer scientists can even contemplate building, it's no wonder that neuroscientists have barely begun to understand how the whole thing works.

Ramón y Cajal also noted that axons, which send electrical impulses, don't actually touch the dendrites, which receive them. They're separated by tiny gaps, known as *synapses,* which are no more than a few ten-thousandths of an inch wide. Somehow, the impulse has to leap that gap. He could think of only two ways that could happen: either in the form of an electrical spark or through the release of a chemical that tells the neuron on the other side of the synapse that it's time to light up. As Kandel was turning his focus to the behavior of single cells, it was clear that the chemical explanation was correct. Neurotransmitters such as acetylcholine, serotonin, epinephrine, and dopamine ferry the instruction from one cell to another that it should turn on (or, alternatively, that it should not turn on).

Kandel knew about all of this when he took up a research position in 1957 at the Laboratory of Neurophysiology at the National Institute of Mental Health, or NIMH, part of the National Institutes of Health, in Washington, D.C. He still wanted to make fundamental discoveries about how the mind works, but he knew, thanks to his conversations with Harry Grundfest at Columbia, that the key to doing so was to start with neurons, just as physicists were studying the nature of matter and energy by trying to understand protons, neutrons, and electrons. And he had progressed, he writes in his autobiography, "from the naïve notion of trying to find the id, ego and superego in the brain to the slightly less vague idea that finding the biological basis of memory might be an effective approach to understanding higher mental processes." He realized, as just about anyone who thinks about the problem realizes, that memory is crucial to how we learn, to how we navigate the world, and to our fundamental sense of who we are.

When he arrived at the lab, it was still unclear where memory lives in the brain, and where it's processed. Shortly after he got there, however, the Scoville and Milner paper appeared. Kandel read the story of H.M. and his surgery and what happened as a result. "The news had a powerful impact on me and many others," he writes. Before that, Kandel could only talk about the biological basis of memory as "slightly less vague" than the biological basis of Freud's id and superego. Now it was vastly less vague. Henry's case didn't inspire Eric Kandel to work on memory, since he'd already chosen that path. But it powerfully reinforced his conviction that this was the right problem to focus on.

At first, he and several colleagues looked at neurons in the hippocampus itself, which was newly identified, thanks to Henry's catastrophic surgery, as the key to forming new memories. They were literally listening for jolts of electricity emitted from neurons in the hippocampus. The technique they used had been developed in the 1920s by a British neuroscientist named Edgar Douglas Adrian. Adrian had learned how to tap into a neuron's electrical activity by inserting tiny electrodes into individual cells. He connected the electrodes to an audio speaker, and when the neuron fired off a burst of electricity, he could hear a distinct popping sound. Among other things, Adrian discovered that the jolts that represent sensory information—sight, sound, pain—don't get more powerful when they're transmitting the information that a light is brighter, a sound louder, the pain more intense. Instead, they pop more frequently.

Kandel had learned how to tune in to the sounds of neural activity in Harry Grundfest's lab at Columbia, making electrodes out of glass so fine that they could be slipped into individual neurons. He experimented with crayfish—an animal Freud himself had studied as a young man, before turning to psychoanalysis. At NIMH, Kandel and his colleagues slid electrodes into the hippocampal neurons of cats. The speakers they'd hooked up popped unmistakably as the cat's hippocampus received input from other parts of the brain.

They didn't learn much, however. Like every other mammal, a cat absorbs an enormous amount of information from all of its senses, then transforms that information into a permanent record it can retrieve later. The scientists could tell that the cats' hippocampi were popping with information-processing electricity, but how the information was actually being stored was impossible to unravel. To understand memory at the level of neurons—to understand the difference between how a neuron looks before a memory is stored and how it looks afterward—they would have to find a much simpler and more primitive nervous system.

After about six months, Kandel settled on *Aplysia californica,* a giant, ocean-dwelling slug that can reach two and a half feet in length and weigh up to fifteen pounds. It's known informally as the sea hare, because its two antennae make it vaguely resemble a rabbit when it's at rest (but you have to squint to convince yourself). "Isn't it beautiful?" Kandel asked me recently, pointing to a large black-and-white photograph of *Aplysia* hanging on the wall of his office at Columbia University. The photograph was indeed striking. It had been taken by one of Kandel's postdoctoral students, in a style reminiscent of the photographer Irving Penn, who took exquisite pictures of trash. The inherent beauty of the slug itself, however, was a matter of taste.

What makes *Aplysia* so appealing scientifically, Kandel said, his back to a huge picture window with a magnificent view of the Hudson River behind him, is its very simple brain. Humans have a hundred billion neurons, more or less; *Aplysia* has only about twenty thousand. Some of those neurons, moreover, are relatively huge. Human brain cells are less than a thousandth of an inch thick. You can see them only through a microscope. *Aplysia* has nerve cells fifty times larger, he said. They're visible to the naked eye, so it's easy to keep track of which neurons connect to which others. Finally, *Aplysia* is capable of learning, which means it has at least a rudimentary form of memory.

Kandel and a postdoctoral fellow named Irving Kupfermann studied *Aplysia* intensively during the mid-1960s, looking for a behavior they could teach it to modify. They wanted to see the physical evidence of that change. They looked at how *Aplysia* feeds, how it squirts jets of ink to confound predators, how it copulates. (*Aplysia* is hermaphroditic: each individual is both male and female, which makes copulation a whole story unto itself.) They rejected all of these. *Aplysia*'s nervous system is organized into eight *ganglia,* or clusters of neurons, and many of its behaviors involve more than one ganglion. Simpler than a human's by far, or even a rat's, but still too complicated. They needed a behavior that engaged neurons in one ganglion only, to give them the best chance of understanding the cellular activity underlying it.

Finally, they settled on the slug's gill-withdrawal reflex. It's a protective instinct that automatically retracts the slug's external breathing apparatus—the gill—when you touch its siphon, a sort of exhaust pipe it uses to expel seawater and body waste. Kandel and Kupfermann set about teaching the slug to change its response to a gill-touch, in two different ways. The first involved habituation. Touch the siphon once, and the gill retracts. Touch it again, same thing. But if you keep poking at it and nothing bad happens, the slug learns to disregard the touch. It works the same way in humans, Kandel said. "If I keep going like this," he said, tapping a teaspoon against his coffee cup, "it bothers the shit out of you for a little while. But after a while, your brain learns to ignore it."

They also did the opposite, teaching the slug to be more sensitive than normal, not less, to a touch on the gill. Their experiment will sound familiar to anyone who's ever heard about Ivan Pavlov and his dogs. Pavlov taught dogs to associate a bell with food; the animals would eventually begin to salivate at the sound of the bell even when no food was present. To sensitize the slug's reflex, Kandel and Kupfermann would touch *Aplysia*'s gill while simultaneously giving it an electric shock. Before long, the slug had learned that a touch meant that an unpleasant shock was coming. Its retraction reflex became far more violent than normal, and it released a squirt of ink as well. As they were teaching the slug to anticipate shock, the scientists used their electrodes to figure out which of the slug's neurons fired into action as it responded to being touched. In the end, they'd mapped out precisely which of the neurons are involved in the withdrawal reflex.

Then they went on to measure the strength of the signal passed from one neuron to the next in these three situations—that is, with no training at all, after habituation, and after sensitization. It turned out that the signal became weaker than normal after the slug was habituated to touch. Its nervous system had learned not to take a touch seriously. The signal was stronger after the slug was sensitized by a shock, the slug having learned that a touch on the siphon was a really bad sign. A specific memory, Kandel and Kupfermann had shown, was simply a change in how strongly the firing of one neuron leads to the firing of others. Santiago Ramón y Cajal had speculated in the 1890s that this might be true. The Canadian neuroscientist Donald Hebb, who was Brenda Milner's grad-school adviser at McGill University and who sent her to work with Wilder Penfield on his research into epilepsy, had reaffirmed the idea in the 1940s. The mantra "cells that fire together wire together," a distillation of Hebb's theory, is familiar to generations of neuroscientists. But it was only theoretical until Kandel and Kupfermann proved it to be a fact.

What actually happens to strengthen the connection, Kandel and a series of collaborators showed during the early 1970s, is that the neurotransmitters, the chemicals that carry firing or cease-firing orders from one neuron to the next, are released more abundantly by the first cell, or detected more sensitively by the next one, or both. If you shock a slug just once, the string of neurons that governs the withdrawal reflex will be strongly connected for only a short time. The slug's dim understanding that a touch means a shock will remain only in its short-term memory. If you repeat the shock over and over—or the shock is especially powerful—the connection is burned in more permanently. The dendrites of the receiving neuron literally grow new receptors to accommodate the extra flood of neurotransmitters.

Of course, a slug is not a human, and a reflex is only the most rudimentary form of memory. The instinct to protect its gills when something touches its siphon has been hardwired into the slug by evolution over millions of years. A slug never learns to do this; the behavior has been preloaded into its nervous system, just as the human instinct to jerk your hand away when you touch something hot was preloaded into yours. But deciding not to protect its gills, or protecting them with unusual vigor, is very much a learned behavior. Thanks largely to Kandel, neuroscientists had begun to understand the nature of learning and memory on the level of individual cells.

What they concluded through later experiments on more sophisticated animals, including monkeys, is that we learn and retain new information in basically the same way that *Aplysia* does. The image of Aline Johnson as a middle-schooler, leaping automatically into my mind as she approached me on the street; the face and voice of my grandmother, who died in 1967; the taste of *cafezinho,* the intensely sweet and bitter Brazilian espresso my parents let me try on a trip to São Paulo when I was eight—I obviously wasn't hardwired with any of these things when I was born. I acquired them through experience. All of them are still in my brain as a series of neurons, linked together by synaptic connections made when those experiences took place. The ones that were particularly meaningful to me, or that were repeated often, or that were reinforced by my bringing them into conscious awareness frequently, became stronger—although I can also remember some things quite vividly for no special reason I can think of.

The neurons involved in specific memories, moreover, aren't permanently reserved for just those memories alone, Kandel said. Each one is wired to many other neurons, not just one. It can be part of many different networks all at once, just as any given road intersection can be part of a thousand crisscrossing routes from a thousand point A's to a thousand point B's.

The sight of Aline walking toward me on the streets of Princeton triggered my memory of her middle school self, which brought up the orchestra teacher and the awful experience of messing up bugle calls in front of the whole school—but if I had let my mind wander, overlapping networks of linked neurons could have led me pretty much anywhere in my cache of memories sooner or later, along a meandering trail of one thing reminding me of something else reminding me of something else. I might have recalled that Aline's friend Susan was the kid sister of my own close friend Jim. That could have taken me to what Susan wrote in my high school yearbook ("You're annoying!") or to the time Jim successfully taught me to drive a stick shift when several other people had failed, or of the time he and I hitched a ride in Yugoslavia with a caravan of Egyptians who were transporting used cars to Cairo for resale, or of the time I played music for his wedding.

Or maybe I would have thought about another Susan, who played with Aline in the cello section. I had a huge crush on her back then. My mother teased me mercilessly. "She has knobby knees," my mother would say, and I can still hear her voice as she said it. Which might have led me to think about yet *another* cello player, and the romantic walk we took on the Princeton campus one spring afternoon. I wore a yellow T-shirt and carried her instrument for her.

We know that these overlapping networks link conscious memories, at least, and that the complexity of the networks is even greater than the number of neurons would suggest.

"There's no question about it," Kandel said. "One neuron can be involved in one memory, and at a very different time the same neuron can participate in another." But it might be even more richly complicated than that. "It's conceivable, although we haven't explored it," he said, "that even though one branch [of a particular neuron] is involved in one kind of memory, another one could be involved in another, entirely different kind"—the conscious memory of the digits in your phone number, say, but also the unconscious memory of what it feels like to ride a bike.

*Chapter 6*

# ARTIST

When Lonni Sue Johnson began designing *New Yorker* ads for Robert Landau's woolens store, it was more than just another commission for her. Landau dreamed of pulling in the kind of people who read the magazine, but Lonni Sue dreamed of drawing for the magazine itself. She didn't want to be a cartoonist. Her work was frequently spiced with humor, but it wasn't the kind that involved punch lines or captions. It was more intellectual than that—more thought-provoking than laugh-provoking. Lonni Sue wanted to draw *New Yorker* covers.

*The New Yorker* is a weekly magazine, so allowing for double issues, it runs forty-seven drawings or paintings on its cover every year. That sounds like a lot, but it gets thousands of submissions, most of them from established artists, many of whom have already produced work for the magazine. The idea of breaking into such an exclusive club was insanely ambitious. Maggi knew that, but also felt that a little insanity was a good thing in an aspiring artist, so she encouraged Lonni Sue to go for it. Maggi had become a fixture in the Princeton art world, and she knew a couple of *New Yorker* cartoonists who lived in town. So one day she took Lonni Sue and Lini, who by now was an undergraduate at Princeton University, and knocked on the door of one of them.

Henry Martin's studio was a tiny, one-room house on a side street near the university. Martin, who now lives in a retirement community in Newtown, Pennsylvania, about ten miles from Princeton, remembers the conversation vividly. Lonni Sue wanted to draw covers for *The New Yorker.* How should she go about doing that? she wanted to know.

It turned out that Martin also wanted to draw *New Yorker* covers. The fact that he was already an established cartoonist with the magazine didn't give him any advantage. He had to submit drawings just like everyone else. To try to improve his chances, Martin had been studying *New Yorker* covers carefully, for a long time. He'd gotten his hands on every issue ever published since the magazine was founded, in 1925, sliced off the cover pages, and had them bound into books. He would pore over them, trying to absorb the essence of what made a successful cover. He'd been doing this for years, and by the time of Lonni Sue's visit, he figured he didn't have what he called "a Chinaman's chance" of breaking the code. He handed Lonni Sue a couple of volumes and wished her good luck.

He also told her that once she had some paintings or drawings to submit, she should show up at the magazine's offices in New York on Wednesdays, when all of the cartoonists and other artists dropped off their work for consideration by the editors. "Take in your portfolio," he said. "Leave it for a week, get your rejection slip, do it again, get another rejection slip. Keep doing that every week, every month, every year, and maybe someday they'll notice you."

That's what she did. She'd run into Martin there, and got to know some of the other artists, both established and undiscovered, who would also show up every Wednesday. She submitted drawing after drawing for both the cover and for "spots"—tiny drawings, not cartoons, which appear randomly throughout the magazine. It took about three years of Wednesdays before the editors held on to some of her spots, meaning they didn't reject them right away. Finally, they actually bought a couple. Lonni Sue was ecstatic. And then, at last, they bought one of her cover paintings—and then another. The magazine bought more than they used, but five of the paintings actually ended up on the cover.

Maggi and Eddie were living in Japan at this point. He'd become director of RCA's Tokyo research laboratory. She spent her days absorbing Japanese art, whose aesthetic would come to inform her own work. When the first *New Yorker* with one of Lonni Sue's paintings on the cover was published (it depicted a stylized Manhattan cityscape), Maggi walked across the street from their apartment and into the Hotel Okura, where she knew the newsstand carried *The New Yorker.* "This dear little Japanese woman handed it to me," she said, "and I couldn't help myself. I told her my daughter had done the cover." Lonni Sue had made it, in a way few commercial artists ever do.

She was now in her late twenties, and her romantic life was also beginning to flourish. She'd done plenty of dating in college and afterward, but hadn't found someone she wanted to be with long-term. Around the same time she began hanging out with Henry Martin, she began dating a Princeton graduate student named . . . Henry Martin. (The cartoonist remembers calling Maggi on the phone a couple of years later. "This is Henry Martin," he said, with an accent that still bore traces of his Kentucky childhood. "No it isn't!" Maggi declared, indignantly, knowing full well that the boyfriend had no such accent.)

The younger Henry Martin was studying composition in Princeton's music department. Lini was an undergraduate in the same department. Like her sister, she'd started out going to college in one place (Brown, in Lini's case) and then transferred to Princeton. When I asked her for more details during one of our conversations, she said, "It's too complicated. The major things were . . . well, who cares?" I cared, naturally. I'd gotten all sorts of fascinating insights into the lives of three of the Johnsons, and everyone who knew them said that all four of them were extraordinary and fascinating people. Of course I wanted to know about Aline's life, especially since she has ended up devoting it to her sister's care. "There's nothing to know," she insisted. It seemed pretty clear that she wasn't trying to hide anything. She just didn't see any point in talking about herself.

While she was at Princeton, Lini would run into Henry Martin frequently at the music building. They got to know each other casually. At some point, Martin remembers, she mentioned to him in passing that he might like to meet her sister. She thought he and Lonni Sue would get along well. Soon afterward, through circumstances he can no longer remember clearly, Henry ended up playing piano in a chamber music ensemble in which Lonni Sue was the violist. They did get along. They began dating. They got serious. This was in the winter of 1977; by the following summer they had moved in together. In 1979, they moved to New York, where they got married.

It was Lonni Sue's idea to make the move. She wanted to be closer to other artists, and especially to potential clients, and once she'd made her mind up, that was it. A decade earlier, she'd decided at the last minute that she would study art instead of history, and that's what she did. After college, she'd decided to try to make it as an artist, despite the long odds against it, and that's what she did. She wanted to do illustrations for *The New Yorker,* where the odds were even longer, and she did that too. Lonni Sue didn't necessarily come across as someone with relentless determination. She spoke quietly and struck many people as withdrawn and shy—more of a listener than a talker, as her fellow teacher and collaborator Bob

Denby told me. She often seemed to fade into the background, especially in large groups of people.

In smaller groups, however, and especially with people she knew well, she would speak more freely. She'd hold her own even when conversations became contentious. So at Lonni Sue's insistence, she and Henry moved to New York. They ended up at the Ansonia, a storied residential hotel on Broadway, on Manhattan's Upper West Side. Babe Ruth had lived at the Ansonia. So had Isaac Bashevis Singer, Igor Stravinsky, Enrico Caruso. So, eventually, would Angelina Jolie and Natalie Portman. When Henry and Lonni Sue moved in, the Ansonia was also home to Plato's Retreat, the infamous swingers' club. It closed soon thereafter.

Lonni Sue's career flourished during her time in New York. Getting her work into *The New Yorker,* which would ultimately buy a dozen or so of her watercolor paintings and put five on the cover of the magazine, led to a deluge of offers from other magazines, newspapers, and corporate clients. She would go on to illustrate dozens of books; provide drawings to the *New York Times,* the *Washington Post,* the *Wall Street Journal, Time, Newsweek,* and *National Geographic;* and design brochures, reports, and advertising material for scores of major corporations. She especially relished the challenge of trying to convey complex ideas through her art. The image itself was just the tip of the iceberg. Tom Hughes, the art director for the Lotus Corporation, one of her major clients (and also the creator of the original Macintosh computer logo, among many other landmark designs), once wrote about her in a Japanese art magazine: "She presents a fantasy so appealing that you're swept into her work without caution and then, within her world, you're challenged by the concepts alive within it." When he worked with Lonni Sue, Hughes said, her ideas were far more valuable than the artwork itself.

It was that way with all of her clients. Lonni Sue loved to brainstorm. She'd have long conversations where she'd try to get herself into the client's mind-set and produce an image that captured what she found there. She was like the master advertising excecutive Don

Draper on *Mad Men,* except that she not only distilled what the client wanted to say into something surprising and compelling but also created the finished artwork.

Her work was often filled with visual puns, but Lonni Sue was also addicted to actual wordplay. Henry Martin (the husband) recalls it as "delightfully nonsensical." Once, he said, while he was still in graduate school, he and a friend were preparing for their German proficiency exams, and the word "ding" came up, which is German for "thing." Lonni Sue, who didn't know German, thought for a moment and said, "Oh, I see, like chicken ding"—a spicy Chinese dish made with peanuts, also known as kung pao chicken. Henry and his friend thought that this was hysterical, although maybe you had to be there. Or maybe you have to like puns, which aren't to everyone's taste. He can think of dozens of moments like this. She can't remember a single one. She can't remember Henry, either. When you show her a photograph of him, the best she can say is that he seems vaguely familiar. She's always surprised when you tell her she was married to this man for a decade. You might also assume it would be unsettling, that it would put at least a small chink in her relentless cheer. How could you be that close to someone for that long and not remember? It doesn't faze her at all, though. Her surprise is more of the "Isn't that interesting?" variety.

Even at her most successful, however, Lonni Sue could never stop worrying. On the surface, she always seemed relaxed and cheerful. Henry and a few close friends knew, however, that underneath her publicly upbeat, cheerful demeanor she had a constant, nervous energy—a relentless drive that kept her perpetually chasing more clients and keeping the ones she had satisfied, combined with an anxiety that never seemed to let up. Early on, Henry remembers, when he'd ask her what was bothering her, she'd say something like, "I'm so nervous about making it professionally. I can't let it go." Later, after she had made it, she would still fret constantly. "I've got so many jobs," she'd tell him. "I worry about making the deadlines."

She always did, though. An art director named Joe Yacinski, who worked for what is now *Kiplinger's Personal Finance* magazine, became aware of her work when his boss commissioned some illustrations from her. Yacinski was young and ambitious himself. He wanted to forge personal connections with artists he might work with in the future. The magazine was based in Washington, D.C., while most of the artists were in New York. "I found myself kind of making excuses to go to the city," he said, "and would kind of just casually throw out, to different artists, 'Oh, can I stop by your studio?,' or whatever." Lonni Sue was one of them. Yacinski insists that he rarely uses the word "awesome"—"I kind of reserve it for God," he says—but he said that meeting Lonni Sue was awesome. She was a ball of fire, a bundle of energy. He couldn't figure out when she slept. "I never met someone as prolific as she was," he said. They quickly became close friends. Yacinski would sometimes sit and talk with Lonni Sue as she worked. She was in the too-much-to-do phase of her professional life at this point, and he was aware of the pressure she was feeling. But she had no qualms about chatting while she raced to finish a project as the FedEx pickup time loomed.

Yacinski's partner, Ron Flemmings, connected with Lonni Sue as well, but at a more visceral level. He's a fine artist rather than a commercial artist—a painter and printmaker. He saw at once that she was more than just an illustrator. Her talent and her aesthetic ran much deeper. Being in her studio, seeing her work everywhere, he said when I visited him and Joe in their home in Alexandria, Virginia, was like Christmas with all of the presents open and scattered around the room. "I fell in love with this person immediately," he said. She was clearly a gifted artist, he saw right away, but it was more than that. She might be anxious about her career, but she was also excited about life in New York.

The two of them would wander the streets together when Lonni Sue had some time in between assignments, dropping into museums and galleries, talking about art and engaging in competitive wordplay, trading puns back and forth. Sometimes they'd sit in her studio, each of them drawing, mostly sitting in silence but occasionally commenting on each other's work. "We were like playmates," Flemmings says. Both men understand why others might have seen her as shy. She didn't try to draw attention to herself, especially with strangers. Even with friends, she tended to listen more than she talked. She silently absorbed the conversation around her—an active listener, her friend Bob Denby had called her. And when she did speak, she didn't try to command attention. "She would often come out with a subtle pun or a double entendre," says Yacinski, "and just sort of let it lie." Her delivery was completely deadpan. "Then she would kind of watch to see if you got it or not." Yacinski still has a book she illustrated, titled *Eat These Words,* a collection of quotations about food and diet. (Examples: "I want nothing to do with natural foods. At my age I need to eat as many preservatives as I can get."—George Burns. "It's okay to be fat. So you're fat. Just be fat and shut up about it."—Roseanne Barr.) She gave them a copy, and inscribed it: "For Joe and Ron with casseroles of love. Lonni Soup." Yet with her closest friends, men and women alike, she had no trouble talking about herself and about her most closely held thoughts and feelings.

Their shared love of music and art was enough to keep Lonni Sue and Henry together as they began to thrive in their careers, she as a commercial artist and he as a composer and musicologist. After ten years together, however, it was no longer enough. According to Henry, it was Lonni Sue who ended the marriage. She told him there were too many incompatibilities between them. Looking back, he said, he's convinced that this wasn't the real reason for the breakup, and that there were actually two reasons. The first was her father's prostate cancer, which led to his death in 1989, and which affected her deeply. The second reason, he believes, was that she felt burned-out professionally. Her clients loved her work, but they all wanted more or less the same kind of whimsical drawing. It was that frustration, Henry thinks, that led her to take a painting class. She fell for the instructor, and eventually bought an apartment for them to live in together.

But that relationship didn't last, either. At the same time, her anxiety over her work continued to grow. At the end of the 1980s, the publishing industry was just beginning to feel the first hint of what would ultimately turn into a crisis for traditional media of all kinds. Leveraged buyouts and hostile takeovers were forcing magazine companies to keep their investors happy by driving up short-term profits. To do that, they had to cut costs wherever possible. They began laying employees off, and instead of assigning photographers and artists to create new images, they bought photographs and illustrations from stock agencies. Clients were still calling, but not quite as often.

Lonni Sue was now forty years old. With both her romantic and her professional lives increasingly out of control, Lonni Sue decided to flee the city. She moved into a weekend house in New Milford, Connecticut, which she and Henry had bought and renovated a few years earlier. Later, she moved to a Revolutionary War–era farmhouse in nearby Sherman, up on a hill, with two barns. She had a sense of peace once she moved up there. She could work without distraction, and when she looked out the window, she saw mostly woods. There was a stream in back, where a great blue heron would occasionally come looking for dinner. She loved flowers; in New York, she'd had a rooftop garden filled with potted plants. In Sherman, she could have a real garden, where she could get her hands deliciously filthy in the fertile earth. Early on, Manhattan had been exciting and invigorating. Joe Yacinski remembers her talking constantly about how much stimulation she got just from walking back and forth to work and the wind blowing her portfolio. Being in New York gave her a creative charge. But things had gotten unpleasant toward the end. She'd had enough.

*Chapter 7*

# FLIGHT TO COOPERSTOWN

Henry Molaison's small-plane flight over Hartford, Connecticut, when he was a boy was such a thrilling experience that it was just one of two memories that survived the surgical assault on his brain (the other was his first and last attempt to try smoking, which made him unforgettably sick). Lonni Sue's first small-plane flight was thrilling as well. It was also terrifying.

The ride came about because she needed to have the house in Sherman painted. She got to talking with the man who came to do it. His name was Duke, and it turned out that he was a private pilot as well as a handyman. Why didn't she come up for a ride? No, impossible, she answered. I don't have time. I've got deadlines. It was true, but Duke told Maggi later that it was also a convenient excuse. She had no problem flying in commercial airliners, but it was pretty obvious to Duke that she was frightened to go up in such a small, flimsy-looking contraption.

The next day, Duke asked again. No, too busy. He asked her the day after that. Eventually, he wore her down. She was genuinely busy, and genuinely nervous, but some part of her must have been intrigued. She gave in, and up they went.

Lonni Sue was frightened the entire time they were in the air, but she also loved it. Looking down at the landscape and the houses from high above—things you can't see so easily from the much higher altitude and through the tiny windows on a commercial airliner—gave her an entirely new point of view, a different way of seeing familiar objects. As an artist who never stopped observing the world around her, this new perspective was exhilarating.

The next time Duke took her up, she was still afraid, but maybe a little less than before. And by the time the painting job was finished, she was so hooked on flying that she knew she had to get her own pilot's license. Duke agreed to be her instructor. Years later, Maggi suggested to Lonni Sue that it must have been a daunting challenge to do her first solo flight, the final step in qualifying as a pilot. No, not really, her daughter answered. "I was prepared," Lonni Sue told her matter-of-factly. "I had practiced enough." She acted in hindsight as though it wasn't such a big thing, but Maggi was pretty sure it was at the time.

Duke kept his plane at Stormville Airport, a small, privately owned airstrip in Dutchess County, on the New York side of the New York–Connecticut border, about ten miles from Lonni Sue's house in Sherman. By the time she'd gotten her license, she was even more in love with flying. She wanted to go up as often as she possibly could. But every time she flew, she had to rent a plane. It was inconvenient and expensive. She really had to get her own.

Her first plane was a Cessna, but soon she coveted another. It was a vintage yellow Piper J-3 Cub, built in 1946 and now sitting partially disassembled in the hangar at Stormville. The J-3, which would be both more challenging and more exhilarating to fly, belonged to the airport's owner, a man in his eighties named Pete O'Brien. O'Brien had been flying forever. Among other things, he had trained women to fly noncombat missions for the military during World War II—they were known as Women Airforce Service Pilots, or WASPs. Plenty of fliers had tried to talk O'Brien into selling the J-3: the model was relatively inexpensive and very reliable. Pilots thought of it as the Model T Ford of airplanes. So far, he'd refused to part with it. But like so many other people had before, O'Brien took an immediate, very strong liking to Lonni Sue. She asked him if she and Duke could buy the Cub together, and he said yes.

They restored and reassembled the Cub, and she ended up buying out Duke's share. She moved the plane to an airport called Sky Acres, in Millbrook, New York, about twenty miles from Sherman. It was bigger than Stormville and had better facilities. That meant there were more pilots around to hang out with and learn from. They welcomed her into the group, and one of them, a flier named Bob Burke, decided he was interested in more than just friendship. "She was a pretty cute lady," Burke said. "I noticed her right away." She wasn't married, and Burke wasn't married, so they began to date. "We did that thing" is how he puts it. Burke also got to know Lonni Sue's family pretty well during that time. They used to fly down to Princeton every so often, and he would take Maggi and Aline up for rides. He and Lonni Sue flew out to Michigan several times as well, where her uncle Dwight had retired. Dwight got a ride, too. The family owned some property out there, an island on a lake in northern Michigan. She had a dream, he said, of flying out, landing on the water, and taxiing right up to her dock. They went back and forth on what kind of plane she'd have to buy, because you couldn't put floats on the J-3. Besides, the plane was much too slow for that kind of trip. Lonni Sue finally gave up on the idea, but only reluctantly. The idea of parking a plane next to where she lived, however, stayed with her.

The romance didn't last long, but she and Burke remained flying friends afterward. The bonds between pilots tend to be strong. They're members of a sort of secret society, a subculture most people don't know much about. Even the most cautious of pilots is at a far higher risk of premature death than most of us. They'll go on casual outings at the drop of a hat that would be real adventures for anyone else. The Sky Acres group—eight or so of them—would meet up on Friday evenings and fly their separate airplanes up the Hudson River about forty minutes to an airport that had a restaurant they liked. They'd park the planes, have dinner, then fly back home. If this seems like an elaborate way to go about having a meal, it is. Pilots talk about the hundred-dollar hamburger. The hamburger itself costs only five dollars; the rest is what it costs to fly to the restaurant and back. Or they'd fly to a grass airstrip, formerly a cow pasture, in Freehold, New York, maintained by a guy named Clem Hoovler and his wife, Rita. The gang would tie their planes down and go swimming in a nearby river. If they were hungry, Clem would let them borrow his car so they could drive to the deli in town.

Lonni Sue also became good friends with a pilot named Karen Henriques. They met not at Sky Acres but at the Old Rhinebeck Aerodrome, a museum in Rhinebeck, New York, also along the Hudson, which has a large collection of vintage airplanes, and which stages airshows during the summer. It was inevitable that Lonni Sue would end up there. She loved flying, but she especially loved old planes. Open-cockpit biplanes from the World War I era were her absolute favorites. She never owned one, but she would take rides in them every chance she got. Henriques loved old planes, too—she owned a one-sixteenth share in a 1946 Aeronca Champion, known to fliers simply as a Champ. She lived in New York City at the time, but she'd go up to Rhinebeck as often as she could on weekends to volunteer at the aerodrome. There aren't many female pilots today, and there were even fewer back then. The men usually welcomed them, but they also saw them as something of a curiosity. The two women gravitated toward each other. "It's just one of those things," Henriques says, "where someone would say, 'Oh, do you know so-

and-so? She's also a pilot. You're going to have to meet.' Lonni Sue and I became fast friends. She's super."

Henriques became part of the group that would fly up to swim in the river in Freehold and borrow Clem's car. As both women grew closer as friends and gained confidence as pilots, they decided they had to go on a more challenging adventure. They would take a cross-country flight, they decided, heading south to the Sun 'n Fun Fly-In, an annual fliers' convention in Lakeland, Florida. They would take the Cessna 150: it was faster and more powerful than the J-3. It was also more comfortable for a long-haul flight.

The women were confident they had the flying skills to make the trip, which they would break up into four or five segments so as not to get too stressed- or worn-out. They made sure the airplane was in excellent mechanical shape. They'd fly only in calm weather with good visibility. And they had two pilots in the cockpit, so if one of them passed out—a very unlikely situation, but you never know—the other could take over the controls. It was about as low-risk as a trip like this can be. Still, it was potentially dangerous. They'd be flying in uncontrolled airspace, which meant that, unlike commercial pilots, they wouldn't be under the supervision of air-traffic controllers. The first day, which Henriques says was "crazy exciting," they flew down the Hudson River, right past New York City and over the Statue of Liberty. Other pilots' voices would crackle through on the radio, letting people know where they were and where they were headed, but the women had to be on constant alert to keep track of it all. If they made a mistake, it could be catastrophic.

"Some helicopter will describe where they are in relation to the city," Henriques says, "so then you have to think, 'Well, where am I in relation to the city? Okay, I'm here, so they must be over there. Do you see them? Anybody see them?'" Not only did the women have to be on constant lookout for other aircraft, but they also had to watch their altitude. If they drifted too high, they'd enter the controlled airspace surrounding JFK airport, or Newark, or LaGuardia, which could get them in big trouble with the FAA. They might even lose their licenses.

It was kind of a thrill, actually—incredibly stressful, Henriques remembers, but in a good way. Once they were past Staten Island and into New Jersey, they had to land for a while, exhausted. This trip was going to be a lot harder than they thought. They'd made a plan for where they would set down at the end of each day, and none of it worked out as they expected. Later on that first day, for example, the weather turned bad, so they put down at Tangier Island, Virginia, in Chesapeake Bay, where a runway had been built by the military during World War II. There was also a bed-and-breakfast, fortunately, since bad weather kept them on Tangier for three days. Neither of the women had ever heard of the island, which was settled by people from Cornwall, in England, in the 1600s. The people who live there still have a marked Cornish flavor to their Virginia accents.

This setback didn't throw them, however, and neither did several others along the way. It was all part of the adventure, and they rolled with it. Lonni Sue had become more anxious than joyous in her professional life, but the exhilarating sense of freedom she felt while up in the air was enough to overcome any trepidation she might feel about a challenging long-distance flight. In the end, they made it to Florida and back in one piece. "For years afterward," Henriques says, "we would always talk about that as the greatest thing we'd ever done. The greatest challenge. Greatest thrill. Greatest time. It was life-changing. Lonni Sue wanted us to write a book about it." They never did get around to the book. Lonni Sue was just too busy. She would rarely turn down illustration work, because she felt, not without justification, that any given client could drop her at a moment's notice. It had already happened, in fact: *The New Yorker* got a new editor, and Lonni Sue's style suddenly wasn't edgy enough. She was still making a good amount of money—fortunately, since owning and maintaining two planes and a property in Connecticut wasn't cheap. But who knew how long it would last? "And then," Henriques says, "she bought that property in Cooperstown."

Most private pilots keep their planes at a small nearby airport. That's what Lonni Sue did at first. But pretty much from the moment she bought the J-3 Cub, she'd begun dreaming about living on a piece of land big enough that she could have her own airstrip. It wasn't that it was such a huge ordeal to drive a half hour to Sky Acres. She simply loved the idea of walking out her back door, getting into the airplane, and taking off. It was more realistic than getting a seaplane for the island in Michigan, anyway. Even so, it would have been difficult to find the right sort of property in Connecticut or New Jersey. Even with a substantial income, the amount of land she'd need would have been prohibitively expensive. So she and Bob Burke, who by now had completed the transition from boyfriend to friend, began looking a little farther from New York City. They flew north on weekends, scouting southern Vermont and New Hampshire and western Massachusetts from the air. They found nothing.

Then Burke suggested they take a look at Cooperstown, in Otsego County, New York, about two hundred miles north-by-northwest of New York City. He liked to take his plane up that way because it was such a nice flight. So they flew to Cooperstown one afternoon. As soon as Lonni Sue saw the area from above, and realized how beautiful it was, she knew this was it. She was determined to find something. It clearly wouldn't be easy: there were hills everywhere, and not a lot of flat space between them. This wasn't the best part of the world to put a runway no matter how big a plot of land you had. But she'd made up her mind. She and Burke started checking in with real-estate agents until they found one who was enthusiastic about helping Lonni Sue solve her problem. In fact, he already had a place in mind that might work. It was a 165-acre farm at the end of a rural road, about ten miles north of town, along the east side of long, slender Otsego Lake. You had to hike up to the farm's high point in order to catch a glimpse of the lake, but once you were up there you were at the second-highest altitude in the entire county. You had an unimpeded view in all directions, with visibility all the way to the Adirondack Mountains, fifty miles to the north.

Best of all, there was a field on the property that could serve as a landing strip. It was free of big rocks, and it was reasonably flat. It sloped uphill, but only slightly. That wouldn't be a problem. She and Burke paced it off. It was about fifteen hundred feet long. That was a lot shorter than Lonni Sue was used to, but Burke assured her that she could learn to be comfortable taking off and landing in that amount of space. The farmhouse was right next to the potential landing strip. The house was, Burke recalls, "not good." It was small and dirty and poorly maintained. But there was also a big equipment shed where she could hangar a plane, and a barn in good structural condition. That wasn't typical in the region, where many barns have collapsed. You could put a decent studio in there, she thought.

It took a few months and several return visits before Lonni Sue was ready to commit. Henry Weil, a physician who owns the adjacent land, was present for one of them. Weil and his wife, Rebecca, don't live on this property. He bought the land for cross-country skiing and grouse hunting. The latter is an indefensible habit, he admits, but he loves it anyway—you're out in nature, tramping around through brambles and brush with a dog at your side, trying to flush the birds out and almost never managing to shoot one. "It's just hard to explain the appeal if you haven't done it," he said, almost apologetically, as we sat in his office at Bassett Hospital.

Weil was out hunting on a day in late fall. He was coming down from the lookout point at the top of the hill. A fierce wind began to howl, and then snow began to fall. The wind blew it nearly horizontal. Above the wind he heard a pickup truck grinding up the road. It was the real-estate agent, bringing Lonni Sue up for a look at the farm. Her eyes widened at the sight of this disreputable-looking person, a shotgun on his shoulder, carrying a bag with bloody feathers sticking out of it. They spoke briefly, and he still remembers vividly how shy she seemed. He also remembers thinking that this was the last that Otsego County would see of her. The terrible weather, the dilapidated farmhouse, the bizarre character emerging from the trees with a gun . . . no way this woman was ever coming back.

But Lonni Sue took possession of that same property in the spring of 1999, the year she turned forty-nine, and just a few months after the encounter with Henry Weil and his shotgun. She'd sold the house in Connecticut, packed up her things, and moved to the place she had renamed Watercolor Farm. The house was still dilapidated, but rather than fix it up, the first thing she did was to start work on the landing strip. She hired a local contractor to remove some big trees at the end of the field, and to clear out brush that had grown up in the years since it had last been used to grow crops. Before long, the grass strip was ready.

Lonni Sue was not.

She'd been terrified to go up in a plane the very first time she was offered a ride, back in Connecticut, and while she'd overcome that fear, she always remained extremely cautious when it came to flying. "Super timid" is how Karen Henriques put it. Bob Burke felt the same. She was a very good pilot, he says. She had strong stick-and-rudder skills and could handle a plane extremely well. But she never had a lot of self-confidence. It didn't help that two or so years into her flying career, she'd had to abort a takeoff and come down in a cornfield. It was her only flying accident, and there was no damage ("except to the corn," Maggi would make sure to add when she told the story afterward). Some people would probably never get into the cockpit again after a close call like that, but Lonni Sue, just a bit more determined than she was fearful, did.

She worried about everything, though. It's too windy to fly, she'd say, or the clouds aren't high enough, or it's a little hazy. I'd better not go. You'd have to talk her into it. "C'mon, Lonni Sue, you flew last week when it was windier than this," Bob would tell her, and she'd answer something like, "I know, but I just don't feel good about it." They'd go through this sort of exercise frequently. In fairness, the landing strip was only about a third the length of a standard runway. Any pilot would have been foolish to try to land there without a lot of experience. Lonni Sue got her practice at a small airport in nearby Westfield, executing progressively shorter takeoffs and landings until she could do both comfortably in under fifteen hundred feet. It took her two years, but she finally brought the J-3 down right next to her farmhouse.

During that time, she also got the house in livable shape, had the machine sheds modified so they'd fully accommodate the Cub, and converted a corner of the barn into a working studio. She also had a backlog of manure from long-departed cows excavated from the barn and used it as a base for flower gardens and a vegetable patch. The farm itself began to take on the colors of a Lonni Sue Johnson painting. She filled the house and the barn with artwork, hung practically edge to edge. Most of it, though not all, was her own. She resumed her daily routine of grinding out one illustration after another, often on very tight deadlines. *The New Yorker* had mostly stopped buying her work, but she still had a biweekly assignment from the *New York Times* to illustrate a column in the paper's business section. She'd have to come up with at least a dozen possible concepts for the editor to choose from, then flesh out the one he picked, often in just a day or two. This was before broadband Internet, so the FedEx driver became a frequent visitor. With her permission, he began timing his deliveries whenever possible so he could spend his lunch break fishing in her pond.

Lonni Sue made friends with the Weils, who owned the property next door but lived several miles away. Henry and Rebecca spent a lot of time walking or skiing their land, or hers, or both. Sometimes they'd take a picnic up to the high point and meet up with Lonni Sue before or after. Both Henry and Rebecca took to her right away, and she and Rebecca quickly became especially close. "She was just one of those gems," Rebecca told me. "I met her and sort of adored her instantly."

Rebecca and Lonni Sue would go on walks, just the two of them, when Lonni Sue didn't have a deadline hanging over her head. They'd talk about nature and art and writing. Rebecca had trained as an occupational therapist, and she'd set up programs to bring nature and animals and the arts into schools, psychiatric units, and nursing homes. She'd written a book about children and nature in the mid-1990s, and was almost finished writing another. When she was done, they agreed, Lonni Sue would illustrate it. But the Weils had two small children, so the writing was going slowly. "It's a sadness for me that it never happened. It would have been amazing to work with her," she said.

Lonni Sue wasn't like other people, the Weils agreed. She wasn't exactly shy, as Henry had thought when he first met her out on the property. That wasn't the right word. More like self-effacing. She didn't draw attention to herself. She was calm, quiet. She had an unassuming, don't-pay-attention-to-me kind of personality. On one level, she came across as what Rebecca described as floaty, vague, imprecise, a sort of poetic, artsy type. That, combined with a sweetness and innocence, gave her something of a hippieish quality. But when she talked about her business and her plans to survive in a weakening illustration market, she was absolutely focused. She was constantly coming up with new ideas for ways to commercialize her art.

Many people who got to know Lonni Sue after she moved to Cooperstown also noted something darker. "Sometimes," Rebecca Weil said, "you get the feeling that there's another layer to someone. You can feel that they've had some profound pain in their lives. I felt that with Lonni Sue." She was clearly a courageous woman. She did things that a lot of women wouldn't feel safe doing—learning to fly despite her fears, and living alone in a tiny farmhouse where she couldn't see the nearest neighbor and where she had to deal with treacherous roads and bitter cold during the winter.

But she also seemed to Rebecca like an injured, fragile soul. Lonni Sue had been strong-willed right from the start, but the backbone Rebecca saw seemed to have been tempered and strengthened in some sort of emotional fire. Rebecca never asked about it, sensing that this was a place Lonni Sue simply didn't want to go. "I really don't know about her previous life," Rebecca said, "because both of us just started from the moment and went forward."

Henry Weil had the same instinct about Lonni Sue. So did Ted Sumner, an artist who'd grown up with her back in Princeton and now lived in Cooperstown. He got the sense that she'd had some very bad experiences at some point, and pushed the memories back inside herself. Ellen Levine, another neighbor and friend who, with her husband, Harry, alternated between their homes in Cooperstown and Princeton, thought so, too. The Levines had first encountered Lonni Sue's artwork decades before, when Ellen served on the board of the Arts Council of Princeton. They still remember the Princeton Poster vividly. They never met back then, but when Ellen joined the board of the Smithy Center for the Arts, in Cooperstown, in the early 2000s, Lonni Sue was already a member. That's when she and Harry realized that the person who would fly her little yellow plane up over the ridge, and over their house, was the same woman. It seemed to Ellen that Lonni Sue was so deeply shy and retiring you could have frightened her off easily. Like the others, Ellen always wondered if there was something in her past that made her that way.

The other thing many people in Cooperstown noticed, but those who knew Lonni Sue as a young woman evidently never did, was her voice. It was extraordinarily quiet, almost strained, they said. You'd often have to listen hard to hear what she was saying. It sounded to Ted Sumner as though her voice was "strangulated inside of her chest." Ellen Levine described it as something like a stutter, but not exactly; her voice just seemed to get caught sometimes. A Cooperstown friend named Pati Grady called it "quivery, like the voice of the NPR talk-show host Diane Rehm," who has a neurological condition called *spasmodic dysphonia.*

Lonni Sue hadn't sounded like this when she was growing up, and she doesn't talk like this today: her voice is rich, strong, full of warmth. It often sounds as though she's on the verge of laughter (which, in fact, she often is). If anyone asked about her strained, halting voice back in Cooperstown, she would tell people that she had the same condition Rehm had. But she told just a few of her closest friends a more disturbing story. Her voice problem wasn't neurological, she said. It was the result of trauma, of both physical and emotional abuse. She referred to the man who had inflicted it as "the monster." Grady says she knows the whole story, but that since Lonni Sue told it to her in confidence, she doesn't feel right passing on what happened or who the monster was. I asked her twice. Another friend, Lynn Marsh, thinks she remembers Lonni Sue saying it was someone in England. But she isn't sure.

I also asked Aline about it. It was a delicate question to put to someone who is so unfailingly protective about her sister. "This is . . ." she began, and then she paused for an uncomfortably long time. "I don't know enough to say anything that's . . . I don't know enough about her life at particular times. I can say what her voice was like, but I hesitate to talk anything about something I don't really know about, especially if it's a negative thing." Aline was clearly aware of her sister's vocal problem, but she wasn't ready to ascribe it to abuse—and she doesn't think it was cured by forgetting. "My sister's voice was very tense," Aline admitted, but she didn't find that surprising, because Lonni Sue lived a high-stress life, what with constant deadlines and the need to find and keep clients. Her strong perfectionist streak probably contributed as well, Aline thought. She also doesn't remember that Lonni Sue's voice cleared up immediately when she lost her memory. "I haven't noticed it so much lately," she said, "but I'd say, maybe . . . come to think of it now her voice is much freer. But I would say it might have taken a few years after her encephalitis."

Her friends in Cooperstown clearly didn't invent someone called "the monster," however, and one of them finally told me more of the story. Her name is Pamela Livingston. She's an artist who ran a gallery in Cooperstown. Lonni Sue sold some of her work there, and she and Livingston developed a close friendship. Like the friendship with Rebecca Weil, this one centered mostly on the women's shared love of nature and of art. "We didn't dwell very much in the past," Livingston said. But they did talk about the abuse, partly because it was something Livingston had experienced in her own life. The monster was a real person, Livingston said. It was someone Lonni Sue had known in New York, although Livingston thought the event itself probably happened after the move to Connecticut. She wasn't sure. He'd abused her emotionally, and at least once, Livingston said, he'd attacked Lonni Sue physically. He'd put his hands around her throat and choked her. She'd been terrified that he was going to kill her. The physical injury had healed relatively quickly, but the memory of that terror had never gone away. It lingered in her mind and also, it seemed, in her throat. Perhaps her voice was constricted because her neck had been constricted, with violence and anger. It testified to the grip of those choking hands every time she spoke.

Whoever the monster was—Livingston either didn't know his name or wouldn't say—Lonni Sue confessed to her and to a few others that she was afraid he would show up in Cooperstown one day. He never did. The move away from Connecticut might have had largely to do with wanting to live where she could taxi up to her back door, as most people believed, and also with wanting to be more in touch with nature. But those few friends who claimed to know the full story were convinced that it also had plenty to do with wanting to retreat to a place of physical and psychological safety.

. . .

Lonni Sue began teaching art classes. She also started writing a column for the local newspaper, the *Freeman's Journal*. She called it "An Aerial Perspective." Unsurprisingly, it usually had something to do with flying. But she also wrote about the farm, her cats, and pretty much anything else that came to mind. She was a natural writer, with a style reminiscent of *The New Yorker*'s E. B. White, who often wrote about life on his own farm in Maine—the inspiration for his best known book, *Charlotte's Web*. Here are excerpts from two of Lonni Sue's columns:

> We turn towards Otsego Lake, a chameleon jewel set between ridges, mirroring the sky. The water is still, except halfway down, where the Southwest breeze scuffs the surface. I'm suspended over the subject of my new paintings. The lake is nine miles long and narrow, clawed out of the deep limestone layer by glaciers. Its presence is overwhelming. Nature rules, man's development still mostly a quiet counterpoint.

> After all of the rain, the giant hollyhocks, double candy pink and midnight maroon, have grown tips too tall to touch. Bow with the heat, nose level yellow lilies open in between a walk up to the barn and back, lacing the air with sweet. Bee balm up to eyebrows spreads in wide red mounds, wreathed with the drum of humming birds. I stop. The height of the garden is so surprising. My sister and mother drove up from New Jersey bringing books, stories, food, and best of all, themselves.

She joined a local orchestra called the Fly Creek Philharmonic. It was what Chris Kjolhede, a pediatrician who played with the group, calls a "tongue-in-cheek orchestra." What he means is clear when you learn that Kjolhede plays the kazoo, the fifty-gallon drum, and the jug. Others blew on disassembled organ pipes or twanged washtub basses, or scraped at washboards, or banged on cowbells. Back in the 1990s, the group did a guest appearance on Garrison Keillor's *Prairie Home Companion* public-radio show. When Lonni Sue joined up, she mostly played the viola, but also took her turn on the pennywhistle and the chamber pot. They'd rehearse all winter and perform once or twice a summer, mostly classical and folk music, keeping up a patter of one-liners in between numbers. One year, Kjolhede remembers, all of the jokes were about trashing the viola or the viola player. "She loved it," he said. "She took it totally in stride." Everyone in the group adored her. She took a year off from the orchestra—nobody remembers why—and when she returned, the other players gave her a standing ovation.

Another Fly Creek Orchestra member, Nicholas Frirsz, was a violinist and a pilot, a kind of male Lonni Sue. Long before they met, he'd been one of the many pilots who had seen the J-3 Cub disassembled and in storage at Stormville and lobbied Pete O'Brien, unsuccessfully, to sell it to him. He was startled when that same plane flew into Westfield airport one day, fully restored. He recognized it immediately. Frirsz was also a violin maker. He would ultimately restore Lonni Sue's viola, and do some work on Lini's cello as well. Along with Bob Burke, he helped Lonni Sue get over her fears about operating out of her tiny, homegrown airstrip. "Nick, talk me through a landing," she would say. "I want to know how it smells. I want to know what it looks like. I want to know what it sounds like." So he did all of that, and then they'd get in the plane, and he'd talk her through the actual landing. He became part of her circle of Cooperstown-area pilot friends, who would meet up at the airport to talk planes and set out on excursions.

By the time she'd been there for two or three years, Lonni Sue had become a familiar sight around town. She mostly kept to the farm, but she'd show up at art openings or at the farmers' market. She was so unobtrusive that it seemed as though she'd always been there, up on the farm, her little yellow plane buzzing overhead every so often. She didn't quite rise to the level of a town character. She was too low-key for that. "There are people who make a mission out of making themselves colorful," Henry Weil said. "It's a total pain in the neck and utterly exhausting to be around." Lonni Sue was eccentric, but it wasn't a conscious thing. She didn't act or dress flamboyantly. It was quite the opposite. One of her art students would tease that she must not own an iron, given how rumpled she always looked. It was only when you got into a conversation with Lonni Sue that you realized you were talking to someone whose way of looking at the world was different from anyone else's.

*Chapter 8*

# BUZZ THE COWBOY

To hear Buzz Stetson tell it, the conversation that led to the organic dairy business at Watercolor Farm happened in an almost cinematic way. He and Lonni Sue had hiked up to the top of the ridge together one day, to the second-highest point in Otsego County, as they often did. Standing on a rock outcropping, framed by panoramic views of the lake to the west and the mountains to the northeast, he gestured to the fields below and said, "You know, this would make an ideal cattle pasture." Then he laid out his vision for a dairy operation that would bring the old farm back to life and provide a stream of income for both of them.

Or maybe it wasn't exactly like that. The film version could legitimately carry the tag "based on a true story," however. The dairy business was Buzz's idea, and he was a very persuasive guy. He is a rough-looking character: tall, skinny, grizzled, haggard, with a respectably thick gray beard. He generally wore torn jeans or overalls and a stained plaid shirt. "I'm pretty much a dirt ball," he told me during a visit to the new organic dairy he'd set up in nearby Richfield Springs after Lonni Sue's illness ended their partnership. If Buzz had walked out of the trees instead of Henry Weil that day when Lonni Sue came out for another look at the property, with a shotgun over his shoulder, she very well might have disappeared and never come back. In fact, he showed up at Watercolor Farm a couple of years after that, as part of a crew she'd hired to install new siding on the barn. They got to talking, and according to Buzz, they hit it off right away. He might be rough around the edges, but, as Rebecca Weil acknowledged, "he can be very chatty, and wonderful, and charming, and warm." Buzz likes to say he'd rather be wise than smart, but if you talk to him for five minutes, it's clear that he's plenty smart.

Buzz Stetson—his given name is Brian—was raised on his father's dairy farm in western Massachusetts. After high school, he went to the agriculture college at the University of Massachusetts, in Amherst. He did well, he said, because he has a photographic memory. "I'd go to class," he said, "take good notes, look at my notes, go to the bar and drink beer before a final exam, clear my brain of all the crap, wake up in the morning and have breakfast, ace the exam. I was lucky." This being the 1960s, he also smoked a lot of pot. Probably too much, he says. "I smoked pot; I had hair down to here." He indicated his hip. "I looked like ZZ Top." To his mother's distress more than his father's, he dropped out just eight credits short of his bachelor's degree and came back to the farm. "I'm a risk taker," he said. "Like it says in the country song: 'I'd rather spend ten seconds in the saddle than a lifetime in the stands.'" He prefers music that has something to say, Buzz said. "I never really was a fan of electric music and boogie," he told me. "Never made a whole lot of sense to me, just a whole lot of screaming." Occasionally, after he'd gotten to know Lonni Sue and her family, he'd sit with Maggi listening to the sisters playing classical music together at the farmhouse. He'd usually fall asleep, he says. It's not clear whether he found the music soothing or simply dull.

Shortly after they met, Lonni Sue offered Buzz a plane ride. He accepted, then rushed home to change into a clean white shirt and penny loafers. "She was surprised by that," he said. For Buzz, this was an occasion like going to church, so he figured that getting himself cleaned up would be the right thing to do. They pushed the J-3 Cub out of the shed by hand. Then Lonni Sue started the engine by yanking down on the propeller, just like Buzz had seen in old black-and-white movies. She had chocks under the wheels—chunks of wood cut at an angle, like enormous doorstops, wedged in front of the tires so that the plane wouldn't start moving before she was ready. The chocks had ropes attached. Once they'd climbed into the plane and were ready to go, she used the ropes to yank the chocks away and begin taxiing slowly over to the landing strip for takeoff. Then up they went, banking south, then west over the ridge, Otsego Lake appearing beneath them as they gained altitude. It was his first small-plane ride. "It's like looking at the world in a flowerpot," he told her afterward. She wasn't entirely sure what that meant, but he clearly enjoyed it.

Before long, Buzz had moved into an old hunting shack on the property, doing maintenance and other odd jobs in exchange for his lodging. He'd lost his own dairy farm not long before, in a divorce, and he was feeling adrift. "I was a country boy, a cowboy, wild and untamable," he said. "I had my own set of issues when I moved to the farm. Lonni Sue always said, 'You're so damn mad all the time. You're so angry. You can't be this angry.'" She was right, he admits. He was angry, about the divorce, and the loss of his farm, and about life in general. He would go on long rants about the unfairness of it all. Lonni Sue would listen. She showed him compassion, he said. "She made me melt."

Sometimes they would have a cup of tea together in the afternoon, and sometimes they'd have dinner as well, followed by a glass or two of wine, sitting in front of the fire and talking. If it got too late, Buzz would sometimes sleep in the spare bedroom. Usually he'd go back to the hunting shack, and then, after a few years, to a small house he began renting farther down the road. He introduced her to some of the neighbors. Lonni Sue was fine at getting to know the artists, and the environmentalists, he says, but he wanted her to know the real people—the guy down the road, the auto mechanic, the tractor-repair guy. He didn't just want her to be the girl who drove by with a wave on her way to someplace else. He introduced Lonni Sue to Esther Hayes, whose family had been farming in this area for more than a century, and to Kay Anichini, the pharmacy technician who would one day rush a feverish Lonni Sue to the hospital.

The truth was that Lonni Sue knew plenty of what Buzz called "real people." She simply wasn't outgoing enough to feel comfortable chatting them up without a reason. Still, while Buzz was a little off base in his assumptions, the introductions helped. Kay's son and daughter would show up at Watercolor Farm, just to visit. Lonni Sue began dropping in on Esther Hayes, asking the older woman's advice about where to buy some geese or how to plant a vegetable garden. Lonni Sue had always wanted to see a calf being born, and Esther had a cow that was ready, so she invited Lonni Sue down to watch. "She was seeing birth for the first time from an animal," Esther said, "and oh my God, it was like you had gave her the whole world." She's a super person, Esther said. "That's all I can tell you. We got along fine. She got along with her neighbors fine."

Rebecca Weil told me that Lonni Sue was easy to fall in love with. She meant it in the broadest sense, but for Buzz it was a little more focused. "I loved her like a sister," he said. "I loved her like a wife. We could finish each other's sentences and things. But we were untainted by the physical stuff." He thought that they might be married for real someday. She held back, he was convinced, because her family had expectations about the kind of man she should be involved with. Henry Martin the composer was an appropriate match. Buzz Stetson the farmer and cowboy and country-music lover and self-described dirt ball, clearly wasn't.

Buzz thought they'd marry, but it's pretty clear that Lonni Sue didn't. For at least part of the time she was living at Watercolor Farm, she had a boyfriend named Roger. He was a commercial airline pilot who would fly his private plane in from his home in Michigan to see her. Buzz never cared for the guy too much. Roger was a pretty boy and a ladies' man, he said. He didn't love her the way she needed to be loved. "I'm sure he slept with her," Buzz said. "It didn't matter to me." Others weren't entirely convinced that it didn't matter. "Yeah. Roger," Kay Anichini said when I brought up his name. "Buzz was just agitated every time he knew he was going to come. He would not show his face the whole time. He would not go in the house."

Lonni Sue made it clear to several of her friends that she didn't share Buzz's romantic ideas about their relationship. She was intrigued with his thoughts about a dairy operation, however. The idea of having a working farm appealed to her. This was at least partly because it seemed to her that this was what the property was meant to do. Pulling the place together and having it operational again matched her own vision. If you have any connection to the land, Rebecca Weil said, it's sad to see a barn rotting. You want it to be vibrant and happy and used. Lonni Sue had that connection. Buzz saw it in the "stick walk" she'd go on every night just before dark. She'd trace the hedgerows that divide one field from another and pick up dry sticks to bring back to start the fire. She always wanted to do it alone. "It was her joy," Buzz said. He thought it gave her a way to be closer to the land. She didn't just want the farm to be a stage set, a scenic backdrop to her life. She wanted to improve it.

That was one reason Lonni Sue agreed to go into the dairy business. The other was financial. The general decline in profitability that had begun to affect the newspaper and magazine businesses before she moved to Cooperstown was getting worse. She had to work harder than before to find new clients, and since they often paid less for artwork than her older clients had, she had to take on more assignments to keep up her income. Lonni Sue's life ambition was to be as successful as Maggi artistically and as successful as Eddie financially. But while Lonni Sue's father had gone into electronics at the very beginning of the computer revolution, Lonni Sue had gone into commercial illustration not long before the field began to wither. Her expenses weren't negligible, either. Owning and maintaining one airplane is so costly that many pilots own only a fraction of one—Karen Henriques, for example, with her one-sixteenth share. Lonni Sue owned two planes outright. If she could have an independent stream of income from a side business, it would take some of the pressure off. Buzz convinced her that the dairy could do that.

So she said yes. Buzz fixed up the barn. He began buying equipment and cows. Lonni Sue named the cows: Lilac, Spatter, Goose, Selena, Brickle. They ultimately had about sixty head, mostly dairy, but also a few beef cattle, and some pigs as well. She handled the books and paid the bills; he handled the cows and the equipment. (The exception was one night in 2006 when Buzz's daughter got married. He got a little tetchy, as he puts it, drinking wine, and came home late. That one time, she helped with the milking.) Within a year, they were shipping milk.

The dairy business turned out to be more problematic, however, and less profitable, than Lonni Sue expected. It's possible that Buzz might have oversold the promise and romance of dairy farming when he presented her with the idea, and downplayed the potential difficulties. "If I wasn't a man of faith I wouldn't be doing what I'm doing," he said. You have to have faith to survive in such a risky enterprise. If you have livestock you're going to have dead stock. Cows die, people die, tractors break. Buzz's hay mower had just broken on the morning of my visit, and this was the second time it had broken in less than a year.

Lonni Sue was temperamentally inclined to take on big challenges. In the case of her artwork and her determination to become a pilot, she succeeded. When it came to the farm, however, she admitted afterward to Maggi that saying yes to Buzz might have been a mistake. Cow smells trickled into her studio, which was built into the barn. The dairy operation turned into a money pit, and the bookkeeping was a nightmare. She and Pati Grady would joke with each other about how running a small business was like being slammed with a financial tsunami, but Pati could tell that it was a lot more difficult and a lot more stressful than Lonni Sue had ever imagined, or was willing to admit. Even a conventional dairy is a huge commitment, and producing organic milk is far more exacting than producing ordinary milk. You can get secondhand machinery for processing the regular kind, but you generally have to buy new machinery for organic. It's extremely expensive. Lonni Sue, she felt, had been a little naïve about the whole thing. The dairy business didn't ease her stress over the art business. It made it worse.

That was how things stood in December 2007. Lonni Sue was worried about her career, worried about the dairy, and very likely worried about her relationship with Buzz, which was never going to be the romance he seemed to want. She considered him a friend. She needed him to take care of the livestock and the farm machinery. But being on the receiving end of constant unrequited love can be draining. She was utterly exhausted. A few days after Christmas, the metal chimney of her wood-burning stove blew down in the middle of a storm. The house began filling with smoke. Lonni Sue called Buzz, then ran outside and climbed up a ladder to hold the pieces in place until he could pull together the tools to make a hasty repair. She was up there in the numbing, freezing, unrelenting wind for the better part of an hour. When she finally got back inside, she took a hot shower and bundled up, but it didn't help. "I can't get warm," she told Maggi on the phone, repeating it over and over. "I can't get warm; I can't get warm."

That might have been the first sign that the herpes virus, dormant for years in some hidden corner of her nervous system, had awakened to prepare its assault on her brain. Nobody knows for sure why HSV1 decides to wake up after years of suspended animation, but doctors think that an immune system weakened by stress and exhaustion might give the virus an opening. The day after the chimney incident, Lonni Sue complained to Buzz about what she thought might be the flu—the worst she'd ever had, she said, accompanied by a blinding headache. A few days later, Lini sent Lonni Sue an e-mail about a drawing Maggi had done. The Arts Council of Princeton was throwing a celebration for their mother's ninetieth birthday. The council asked Maggi herself to design the invitations, and she and Lini wanted to know what Lonni Sue thought. They got some e-mails back, but they were, as Lini would later remember, "a little flabby and short. But as the younger sister," she said, "I didn't always get the most attention."

Friends who saw Lonni Sue or talked to her during that period thought she was acting strangely. Lynn Marsh, whose farm sat down at the bottom of Lonni Sue's road, stopped to see her about a week before New Year's Eve. She wanted to discuss a painting she'd commissioned as a surprise anniversary present for her husband. Lonni Sue had done several sketches, and Lynn had finally decided which one she liked best. To her surprise, Lonni Sue had taken down all of the artwork that decorated the walls, and couldn't really explain why.

During a typical visit, the women would sit at the kitchen table and have a cup of tea and talk for a while. This time, Lonni Sue was scattered. She didn't want to discuss the painting, or even have anything to do with it. She couldn't seem to figure out how to make the tea. She didn't want Lynn to stay. "She was actually cruel," Lynn said. "She'd never been that way before, ever."

At about the same time, Joe Yacinski got a voicemail from Lonni Sue. It didn't feel quite right. It wasn't the warm, embracing kind of message he was used to hearing. It seemed distant, minimal. He tried calling back, but got her voicemail in return. It never occurred to him that anything was seriously wrong, so he didn't pursue it. "I mean, now I'm, like, 'Oh my God, early warning signal, I should have done something,'" he said. Months later, after Aline managed to get into Lonni Sue's e-mail account, she could see that her sister had been in constant contact with hordes of people, writing e-mails that Aline describes as full of "florid paragraphs, and inspirations, and things that she was talking about with her friends." And then, about halfway through December, it dwindled to nothing.

During the nine years Lonni Sue had been in Cooperstown, her mother and sister would often drive up to spend the Christmas holidays at Watercolor Farm. The year before Lonni Sue got sick, however, they'd gotten caught in a blizzard. They didn't want to take the chance of that happening again, so they decided to skip it this time. If they'd been there, Lonni Sue's strange behavior would have been impossible to miss. Maybe they would have gotten her to the hospital sooner so the doctors could have stopped the virus from doing so much damage. Even a day or two might have made a difference. Maggi and Aline didn't dwell on this afterward. There was no point. But they couldn't help wondering.

In the end, they didn't find out that something had gone terribly wrong until the telephone call came.

*Chapter 9*

# FIRST TURNING POINT

The phone rang at seven o'clock in the morning on December 31, 2007, in the house that Maggi and Ed Johnson had built in 1951. Maggi had been living there ever since, except for the years when she and Eddie were in Tokyo and a year they spent living in Switzerland. Aline had moved out years before, but she would sometimes spend the night to keep her mother company. She had slept over the night before. The jangling jolted Maggi out of a deep sleep. She and Aline both tended to stay up long past midnight, then sleep late. A phone call at that time of the morning was very unusual. It was certainly much too early for anyone to be wishing them a happy New Year.

When Maggi got off the phone, she woke Aline and told her

that something awful had happened. Buzz Stetson had just called, she said. He was at Bassett Hospital, where Kay Anichini and her daughter, Maya, had taken Lonni Sue the evening before. Buzz had stayed behind to milk the cows, but he got down to the hospital as soon as he could, driving through what had turned into a raging blizzard. He'd been there all night. The doctors didn't know what was wrong, but it was serious. Buzz had given Maggi the number of the resident on call at the hospital so she could get more information. She phoned, and the doctor filled in all of the available details: Lonni Sue had been admitted the night before with a fever of 104 degrees and severe confusion. They were doing blood tests. They'd also done an MRI and a spinal tap, but the results weren't in yet. Her symptoms were consistent with either bacterial meningitis or viral encephalitis, so she'd been put on both intravenous antibiotics and intravenous antiviral medications, just in case. She'd been sedated and restrained, the latter because in her confusion she kept trying to pull out her IV needle. Early in the evening she still appeared to recognize Kay and Maya and Buzz. As the night wore on, she no longer could. She had been in critical condition when she arrived at Bassett. Now she was worse. The blizzard of the previous evening had finally passed. The roads were still in pretty bad shape, but Maggi and Aline had begun packing for what they knew could be a long stay as soon as they'd gotten the news that morning. The moment they got off the phone with the doctor, they set out for Cooperstown.

They drove off, Aline at the wheel. Ordinarily, the trip from Princeton to Cooperstown takes about five hours. This time, due to driving conditions that became increasingly treacherous as they penetrated deeper into New York State, it took much longer, although neither Maggi nor Aline could say afterward just how long. They couldn't remember what they talked about in the car, or even whether they talked at all. Darkness came on before they were halfway there. All Aline can remember is that she was driving, "like a laser beam," she said, the oncoming headlights streaking by, with an absolute focus on the mission of getting to Lonni Sue. Nothing else mattered. Nothing else existed except for her sister, who lay in a hospital bed somewhere in the darkness ahead.

They finally arrived shortly before midnight—Aline remembers looking at the clock in Lonni Sue's room as they walked in, for some reason. She can still see it if she closes her eyes. The second thing Aline noticed was that although the woman lying strapped to the hospital bed had Lonni Sue's features, she was barely recognizable. Lonni Sue was asleep, but even so, her essence seemed to be gone. It was as though someone had turned a tap and let the personality drain out.

Aline and Maggi slept fitfully in the hospital room that night, half sitting up in reclining chairs. When Lonni Sue woke up the next morning, she seemed baffled, much as Buzz had described her after he broke into her house a couple of days earlier. She would look around the room with her mouth wide open, with an expression on her face of what seemed to Maggi like astonishment and awe. It was as though her brain couldn't comprehend what her eyes were seeing. Every so often, she gazed at her hands—first the palm of her right hand, then the palm of her left. Then she turned her left hand over and stared at the back of it for a while. Then she returned to the right. She seemed mystified, as though thinking, "What are these things? Do they have something to do with me?" It reminded Maggi of how Lonni Sue had been in the delivery room at Princeton Hospital on the day she was born. Her mind had been a blank slate. It seemed to have been wiped clean once again.

The next day she seemed better. She was able to sit up in a wheelchair, and when Aline handed her a pen and a notebook, she began drawing—not figures, but boxes, lined up in rows. As she drew, a jumble of words came out of her mouth, but they didn't make any sense. "I heard one coherent string," Aline told me. "She said, 'I think we need to have a door.'" Aline had no idea what it meant. "I love you," Aline said. Lonni Sue responded, "So my." Maybe it meant "So am I." Maybe not.

Later that day, Lonni Sue went downhill again. She couldn't sit up. She stopped speaking. She couldn't move her limbs. She was still confused. She wouldn't pick up a pen and draw again on her own initiative for nearly a year. Her fever persisted—it would get better for a few hours, but then it would spike again. EEG readings suggested that she was having mild seizures, which is never a good sign. The doctor told Aline and Maggi, "You count the number of good days, and you count the number of bad days, and you sort of see where you are with that." Maggi was desperate to be helpful. At one point, she felt cold and, noticing that Lonni Sue was uncovered, figured she must be cold, too. She tucked a blanket around her daughter. A nurse came in and took it off. "That's no good; she has a fever," the nurse said. "We didn't know how to behave, I guess," Maggi said.

Aline tried to be helpful in her own way. She asked millions of questions, in a respectful but relentless way. She took meticulous notes about her sister's demeanor, her medications, the slightest hint of a change in her condition, for better or worse. When the nurses and doctors went off shift, she and Maggi would brief the next group on what had happened over the previous eight hours, and ask a million new questions. She was determined not to miss even the tiniest detail that might help the doctors figure out what was going on, and to make a prognosis.

When the two women had first set off on the long drive to Cooperstown, they had no idea how long they might be away from home. During those first couple of days at Lonni Sue's bedside, they'd come to understand how truly ill she was. Lonni Sue had terrible trouble initiating any movement at all, for example, even the most mundane and familiar—a condition that would persist for weeks. The hospital aides would bring her a tray of food, and she'd just look at it, not knowing what to do. Aline would have to put her sister's hand on the fork, then guide the hand to scoop up some food, then lift it to Lonni Sue's mouth. Once it was in her mouth, Lonni Sue finally knew what was expected. In a situation like this,

Aline and Maggi didn't have the mental space to think about what might lie ahead. It finally hit them hard two weeks later when they took a break and drove up to the farm. Every time they'd driven up this road before, they knew Lonni Sue would be out on the porch to greet them, usually surrounded by cats, with a huge smile on her face. It struck them as they came around the last curve that it wouldn't be happening this time.

Aline also began to think about the future. Normally, when someone is sick, you have a frame of reference for how long it's going to last. A cold will knock them out for a day or two, a bad flu for maybe a week. A major operation might keep someone out of commission for six weeks, or something like that. At first, she and Maggi were thinking that Lonni Sue would be getting better soon, that recovery was just around the corner. But it was already becoming clear that the corner was farther away than they'd first imagined. It would continue to seem farther and farther, and suddenly, Aline says now, "we started to see the magnitude of things." She'd insisted that the doctors be completely honest with them. She and Maggi just had to know. How many cases had the infectious-disease specialist seen of HSV encephalitis? About fifteen in a thirty-year career, the doctor answered. And what were the outcomes? The answer was that a significant percentage had died very quickly. A few had recovered fully. A few had survived, but many had major, permanent neurological impairments, including severe memory loss, intellectual disability, epilepsy, lack of muscular coordination, difficulty seeing, hearing, or speaking, and major personality changes.

Although Lonni Sue was probably no longer in danger of dying, it was still too early to know what sorts of ongoing problems she might have. She might recover her ability to walk and talk, since these problems might be a result of the temporary swelling of her brain as she fought the virus—or she might not. Since she couldn't speak at first, there was no way to test her memory. It was likely, the doctor said, that she'd come away with some level of memory impairment, but it was impossible to say for sure how severe this would be. As Aline processed this information, she began to understand that her sister might never be able to function on her own again. Maggi was ninety years old. She was impressively healthy for someone her age, but she was starting to get frail, and she wouldn't be around forever. It would fall on Lini to be her sister's caretaker, possibly for months but maybe for years, and maybe for the rest of their lives. The two girls hadn't been very close as children, and they weren't especially close as adults, either. Nevertheless, Aline would tell me later, "I want to live in a world where people take care of each other."

It would be an exhausting job, but things could have been worse. Aline wasn't married (although, she points out, a husband might have been helpful). She had no small children. She didn't have a day job. She'd spent more than twenty years as a computer programmer analyst for Princeton University, but she'd left that position well before Lonni Sue got sick in order to work with her mother on a book about art, focused on the principles of visual language Maggi had learned decades before from Josef Albers and from the industrial designer Alexander Kostellow, and which she had taught to many generations of students, including her own daughters.

Aline hadn't ever intended to go into computers. She'd been headed for a career in music, starting with a bachelor's degree from Princeton University. Was it in music performance? I asked. "No, I focused on analysis." This wasn't simply music theory. It was analogous, she explained, to Albers's ideas about how we're drawn into a piece of art and how we navigate it—how contrasts and similiarities and themes and placement of the visual elements of a piece engage our attention. "You can ask the same questions about music," she said, "or about storytelling or sculpture or architecture." Albers's ideas had transformed her mother's life, she said, and went on to have a powerful influence on her own and on Lonni Sue's.

After she got her bachelor's from Princeton, Aline attended the Juilliard School, in New York, to study cello in a serious way. Only the most talented musicians get into Juilliard, which makes it clear just how extraordinarily good Lini was. She hadn't been there very long, however, when she pinched a nerve rather severely. There was no way she could continue practicing and playing such a physically demanding instrument with the intensity required of a professional musician, so she decided to withdraw.

It was, she now believes, a blessing in disguise. "A lot of my friends in the music department at Princeton were composing on the computer," she said. This was back in the 1970s. "It was very cutting-edge at that time. I had always thought that someday I wanted to learn about computers. I sort of marked that in my mind." Now seemed like a good time to get on with it. She wangled herself an entry-level position in a small software company, she said, "with the help of my parents' friends' son's friend's friend. That's how I got started." She learned programming on the job, then took some time off to enroll in a few computer-science courses at Princeton. While she was doing that, she noticed an ad, tacked to a bulletin board at the university, for a programmer analyst job in the Management Information Services department at Princeton. She got the job.

A programmer analyst, Aline explained, does much more than simply writing code. In her case, it meant that she designed and wrote custom software for the university's business-related functions, specifically for the department that handled incoming revenue from tuition and other sources. "They would say, 'I need to do X,' and I'd say, 'Well, how do you do it now?'" she said. Aline discovered that the skills she'd developed in analyzing music laid a foundation for her work on computer systems. She'd come at the problem from one direction, then from another. Then she'd go back and think some more, and figure out what a software solution might look like. It might involve starting from scratch or it might require the modification of an existing program. "There are many different solutions, probably some better than others," she said. "I'd come back with sort of a proposal," she said. "We'd talk about it and they'd say this or this. So we'd go back and forth. That's the analysis part. It was fascinating. I loved it."

In fact, this was very much like the way Lonni Sue dealt with her own clients. The *New York Times* might be running a business column about productivity, say, or customer service, or organizational behavior. Lonni Sue would read the text and think about how to distill the central concept into an illustration. She would come up with a solution—or rather, several: she would usually prepare a handful of alternate drawings, just as she had for Robert and Barbara Landau's birth announcement back in Princeton, and let the client choose. The sisters were very different in personality, with completely unrelated career paths, but they were remarkably similar in the way they went about their creative work.

Aline stayed at her job until Y2K. This episode faded so quickly from the public consciousness after the year 2000 that it feels nearly as antique as Fred Astaire in a tux, the Nehru jacket, or big eighties hair. Y, as you might remember, stood for "year" and 2K for "two thousand"—the turn of the century and of the millennium. It was also the year when, according to some predictions, the world's electronic infrastructure would fail, the electric grid would go down, planes would fall from the sky, and a long list of other catastrophes would ensue. This would happen because programmers in the 1960s had created systems in which dates were recorded in mm/dd/yy format. It would have taken up too much precious memory to put a "19" in front of the year, and it was obvious what century we were in. Someday, the programmers would have to deal with a new century, but the year 2000 seemed impossibly far in the future. Surely, the software would be rewritten before then. But as the turn of the new century finally did approach, it still hadn't happened. This meant that at midnight on New Year's Eve 1999, computer clocks would roll over to the year 00, which the computers would assume was 1900. Flummoxed by time moving backward, went the theory, computers and everything they controlled, which was to say, pretty much everything, would go immediately haywire.

In the end, midnight came and went and nothing happened. Y2K became a punch line on late-night talk shows, for about a week, before it became completely forgotten. The whole thing had evidently been a bunch of foolishness over a nonexistent problem. In late January 2000, I remember seeing a carton labeled "Emergency Y2K Candles" sitting forlornly in the discount aisle of the local supermarket. Mention Y2K to anyone who lived through it and they'll probably just roll their eyes.

Talk to software engineers or computer programmer analysts, however, and you'll get a different story. The only reason Y2K was a nonevent, they'll tell you, is that they and their colleagues worked feverishly for several years to head off the problem before it became a catastrophe. "It was a monumental project," Aline said. It was three and a half years of work, and it was like having two jobs at once, she said—maintaining the old software while preparing for new software to take its place. The new software would be able to turn over without a glitch for centuries, but it hadn't been written in-house; it had been purchased from outside vendors, which meant programmers couldn't tinker with the code. The manual transmission had been replaced with automatic, and for those who preferred more direct control over the machinery, it just wasn't as much fun anymore. "It was a different era now," she said. "And I was just burned out."

At about that time, she happened to read about some neuroscience research going on at Princeton. She was intrigued. "I've got to find out about that," she thought, and so, being a Johnson, she signed up to audit a course. "It was just fascinating," she said, "so I thought, 'While I'm at it, why don't I take a few more?'" In the end, she thinks she took about a dozen courses. After auditing the first few, she began taking them for credit. "I just had to learn more," she said, "so I wanted to do the lab work. I did the papers. I took the exams. I still have my textbooks and my notes. Let me show you." We were talking in the Johnsons' living room, so she took me back through the dining room and into the study, where she had them lined up on a bookshelf, meticulously organized. She pulled one textbook from the shelf. "If I ever had to go on a desert island

I would take this book with me," she said. Princeton doesn't offer summer classes, so rather than sit around without learning for a whole three months, she went up the road to Rutgers to take a summer course on statistics and research. "I was really nuts about this," she said, still holding the textbook. "I still am."

One of the courses she took was titled Cognitive Neuroscience, taught by a German-born professor named Sabine Kastner, who would ultimately become far more important to the Johnsons than any of them could imagine. It was most likely here, Aline believes, that she learned about Henry Molaison, who was known to the world only as H.M. at the time. She couldn't have conceived back then that what happened to Molaison would happen to her own sister just a few years later. "People tend to say, 'Oh, it's so mysterious that you studied neuroscience and then this happened,'" she said later. "They look at it in this mystical way. The way I look at it is that at every moment in your life, you're drawing from every experience you've ever had."

Well, yes. But in this case, Aline was able to draw from her knowledge of neuroscience to a degree that few relatives of amnesia victims had ever been able to do. She knew the role of the medial temporal lobe in general, and of the hippocampus in particular, in the formation of new memories. She knew how the loss of those structures had devastated Henry's life. She didn't know, in those early days, just how much damage her sister had sustained, and she was too consumed at first with worry about Lonni Sue to think much about it. She did have the presence of mind to know that it would be important to monitor Lonni Sue's cognitive state very closely, however. Keeping track of her sister's mental function and how it changed, day by day or even hour by hour, could prove invaluable to the doctors in understanding what was happening deep inside her brain. It wasn't easy to do at first, since Lonni Sue appeared mostly disconnected from and baffled by the world around her. Finally after several days of unresponsiveness, as one of the hospital aides was giving her a bath, Lonni Sue raised her eyebrows—"like this," Aline says, demonstrating in exaggerated fashion, whenever she tells the story. It was so surprising, and her eyes seemed so suddenly alive after days of emptiness, that everyone in the room burst into laughter.

A day or two later, Lonni Sue's shoulders, which had been frozen, came alive. She could move them, and she could turn her head for the first time. "She could look around," says Aline, "like this," demonstrating again. After another day or two, her waist unfroze. She could sit up. The nurses got her into a wheelchair. It was about ten days into the ordeal when Maggi pointed to Aline, and asked, "Who's this?" Lonni Sue looked at her sister and said, "Aline." It was odd because she never used Aline's full name. It was always "Lini." But it hardly mattered. That single word seemed like a miracle.

Maggi was already hoping for another miracle. She had been there when Lonni Sue first began to blossom as an artist. She had been there when her daughter declared that she was abandoning the study of history to focus on art. She'd helped Lonni Sue develop as a professional artist, and helped her get her foot in the door at *The New Yorker*. She knew how much it meant to her to make art, and she and Aline both wanted Lonni Sue to make art again. So shortly after Lonni Sue regained the use of her hand, Aline handed her a pencil and tried to get her sister to pick up where she'd left off a couple of weeks earlier. Lonni Sue could hold the pencil, albeit awkwardly, but that was all she could manage. "Lonni Sue, draw a line, just a simple line," Aline said. She couldn't do it. "The pencil hovered," Aline said, "but she couldn't get it down to the page. She couldn't make it happen."

After sixteen days at Bassett, the acute phase of Lonni Sue's illness had finally passed. Her fever dropped to the point where it was no longer life-threatening, and she could sit up in the wheelchair for several hours at a time. She seemed to have a terrible headache, which wouldn't go away for many months. But the infection had gone into retreat, beaten back under the combined assault of Acyclovir and Lonni Sue's own immune system. Since she was no longer as severely ill as she had been, the doctors felt she could be moved to a rehabilitation hospital. She was in terrible shape, however. She couldn't walk. She couldn't feed herself. She could barely talk, and what she said was only in the ballpark of making sense at best.

It wasn't clear that she'd ever walk or talk or make sense again, either, but if it was going to happen, she'd need intensive therapy. One option was for Maggi and Aline to bring her back to New Jersey so she'd be close to where they lived. After the horse-jumping accident that severed his spinal cord, the actor Christopher Reeve, who also grew up in Princeton and was about the same age as Lonni Sue, went to the Kessler Rehabilitation Center, in northern New Jersey.

There were also good reasons to keep her in New York State, however. Most of Lonni Sue's friends were there. Aline and her mother thought it was important for them to be able to visit her, on the reasonable theory that seeing people she had cared about and who cared about her would be one of the best forms of therapy she could have. It was also clear to Aline, if not necessarily to Buzz, that she'd have to step in, at least temporarily, to help take care of the farm. Henry Weil found an acute rehab hospital in Schenectady, about sixty miles from Cooperstown, where Lonni Sue was sent to try to regain rudimentary control over her body, and to begin the process of learning to talk and feed herself.

She made some progress there—she learned to walk, although only with help, and her speech improved some. Maggi and Aline would encourage her, pointing to a lamp or a chair and saying "What is this called, Lonni Sue? How about this?" At first she'd mostly resorted to stock phrases, such as "I love you," which she would use over and over. Here and there, she said a short sentence. But the doctors in Cooperstown had told Maggi and Aline that short phrases weren't necessarily a meaningful sign of recovery. Keep listening for sentences of eight or more words, they said. "We kept counting and counting," Aline said. By the time she left the acute rehab, the count had gone up over eight. She was talking in what sounded like sentences—except they had no nouns. They were, Aline admits, mostly gibberish. Lonni Sue spoke them with great intensity, however, as if she knew perfectly well what she was trying to get across. "That sounds pretty interesting, and she's very engaged," Aline would think, "but I have no idea what she's talking about."

When the nouns finally did come, they were nonspecific. She used the word "thing" a lot. She couldn't identify colors, which was especially dispiriting for her mother and sister, given how important color had been to Lonni Sue as an artist. When Aline or Maggi or a nurse would point to a marker and ask "What color?," she might say purple. She might say black. Maggi and Aline had been coaching her every day on vocabulary. It didn't seem to be doing a lot of good.

At the end of two weeks—four weeks after she'd first been admitted to Bassett—the therapists concluded that Lonni Sue wasn't likely to improve much more, but that she needed to continue with therapy to reinforce the gains she'd already made. They sent the Johnsons off to another rehab, in Stamford, New York, which was geared to patients who weren't so acutely disabled. This one was about the same distance from Cooperstown as Schenectady, but to the southeast rather than the east.

At this point, Maggi and Aline felt that they could finally breathe a little. The mail was piling up at home, they had bills to pay, and they wanted to vote in the 2008 presidential primary. They jumped into the car and drove to Princeton. They got there just two hours before the polls closed. A few days later, they were back in Stamford, still not knowing exactly what to expect. Every time Lonni Sue went to a different facility, Aline and her mother had to figure out where they would stay. A hotel? An apartment? They had no idea how long Lonni Sue would be there, and they never knew she was moving on until the day before she did. During the acute phase of the illness, while Lonni Sue was still at Bassett Hospital, they could have stayed at the farm in Cooperstown, but the upper end of Lonni Sue's road was choked with snowdrifts and difficult to navigate. As it turned out, the hospital had a guesthouse right on the grounds—a godsend, Aline called it—so they slept there instead. In Schenectady, they slept in Lonni Sue's room, or in a guestroom at the hospital, or in a hotel.

In the third facility, they stayed at the Holiday Inn when they were in town. "Stamford was a tiny little town," Maggi said, "where most of the places were boarded up. It had once been a resort area for rich people, but that was a long time ago." They'd dropped Lonni Sue off and then had to leave right away. "We hoped it was an okay place," Aline said. "We didn't really know."

By now, Aline and Maggi were alternating ten days or so at home in Princeton, tending to their lives there, with a week upstate. The days in New York fell into a predictable rhythm. They'd wake up in the morning, deal with the endless issues that kept arising about the farm or about Lonni Sue's care or her medical insurance. Then they'd head over to the rehab to spend the day with Lonni Sue. She was getting physical therapy, where specialists worked with her to gain a measure of control over her muscles and build up her strength and balance so she could begin to walk. The hospital staff also gave her occupational therapy, which in the jargon of rehab doesn't mean training for a job; it means preparing for the occupation of day-to-day life—making toast, boiling water, putting on your socks, feeding yourself, bathing yourself, getting dressed. And she had cognitive therapy, which involved exercises that helped her continue to associate words with the things they described, and put them together to form coherent sentences.

In Schenectady, Lonni Sue had had three one-hour therapy sessions every day, but when they were over, Aline and Maggi continued pushing her to practice what she'd learned. They would take her up and down the hall, helping her reinforce her training on how to walk. Lonni Sue wore a thick belt around her waist, called a *gait belt*. Aline would grip it firmly from behind, prepared to steady her sister if she began to topple over. They played a game where Aline or Maggi would point to a doorknob, for example, and say, "Doorknob!" The idea was that Lonni Sue would repeat it, but for the first few weeks she wouldn't. Aline tried to get her sister to point to the doorknob herself if she couldn't say the word. "I thought I could at least train her to communicate by gestures," she says. It didn't work.

Once Lonni Sue got to Stamford, her therapy was reduced to three half-hour blocks per day. This was unfortunate, says Aline, because her fever had by now dropped to the point where Lonni Sue was feeling much less ill, and could have made better use of longer sessions. Then something happened that made Aline's jaw drop—literally, she insists. It was February 14, 2008, Valentine's Day, about six weeks after Lonni Sue was first admitted to Bassett. Lonni Sue was sitting in her room in Stamford, looking out the window, and said either "bird" or "bird on a tree"—Aline isn't sure which anymore. Aline looked, and there was, in fact, a bird on a tree. You or I would probably have praised Lonni Sue effusively, felt good about it, and moved on. Aline and Maggi didn't move on. "I wanted her to draw the bird and to talk about it," Aline said, "so first I made a drawing, and I pointed to the different parts—the beak, the head, and so on. Then I told her to draw."

Lonni Sue couldn't make even a simple line while she was at Bassett, aside from that brief window right at the beginning, and despite Maggi's daily urging once she got to Schenectady, she hadn't made any progress. Now, in what seemed to be an abrupt leap forward, perhaps as the result of the drop in fever, she suddenly could. Aline still has her own drawing, and also the one Lonni Sue made in response. "Copying is normally illegal in our family," Aline said, "but we made an exception in this case." And it wasn't really a copy anyway. "She made it more of a bird than you did, Lini," Maggi pointed out later. "It had more character. Do you see how the legs go into the body? It has just a little of that Lonni Sue whimsy. And look, that's the same kind of horizon line she often used in her illustrations." Six weeks after her brain had nearly been destroyed by encephalitis, Lonni Sue was showing a spark, if only a tiny spark, of creativity. The bird offered some hope that the woman Maggi and Aline barely recognized that first night at Bassett Hospital might be gradually becoming herself again.

Maggi and Aline weren't the only ones who were giving Lonni Sue extracurricular help. For some reason, two aides who worked at Stamford were powerfully drawn to her. She was easy to fall in love with, Rebecca Weil said, and it was proving to be true once again, even though very little of the woman Rebecca knew had emerged from the fog of encephalitis. "They [the two aides] knew she'd gotten a poor prognosis from the previous rehab," Aline says, "and they wanted to prove it was wrong." In the evenings, after their workday was done, these women would come back into the room armed with workbooks and exercises. They also brought coins: one of them had been an elementary school teacher, and she was determined to teach Lonni Sue about money, and about how to add and subtract. They didn't waste a minute. "When we saw them coming through the door," Aline says, "the looks on their faces said, 'Get out of the way.' It was time for us to leave. They meant business."

There was also a steady stream of visitors, which had begun even before Lonni Sue was moved out of the intensive-care ward at Bassett. "They came out of the woodwork," Maggi said. "I was astounded that she had so many friends. I thought she was too busy." They came at night and sang to her. They came and prayed for her. One of Lonni Sue's pilot friends set up an easel in her room with the painting she'd done of his plane. He thought she'd be comforted by seeing an airplane, and by seeing her own artwork. Nicholas Frirsz, the violin maker, musican, and pilot who had talked Lonni Sue though her first landing at the strip at Watercolor Farm, brought a guitar and played a tune for her. He can't remember which one. "She seemed . . . well, obviously she was very ill," he said, "but she was somewhat alert and conscious. I played her a little bit of music, and she responded to that almost immediately, where she hadn't responded to anything else before." It wasn't much of a response:

Lonni Sue looked over at Frirsz and smiled, but she showed no sign that she recognized him.

All told, Aline estimates that maybe a hundred people came to visit Lonni Sue at least once over the months after she got ill—doctors, artists, farmers, pilots, businesspeople, musicians, neighbors. Aline and Maggi had seen Lonni Sue frequently when she lived in New York City, but the visits came less often when she moved to Connecticut. When she moved even farther away, they came only a few times a year. Maggi and Aline had no real understanding of her life in Cooperstown. Now they were beginning to. "We were learning about these different facets of Lonni Sue from her friends' perspectives," Aline said, "and it was just a fascinating thing to see." It was also a strange experience. "There she was, especially in the early days, lying in bed, and the conversations were all about her, and she was the only one not participating."

The visitors kept coming when Lonni Sue moved to the acute rehab in Schenectady, and when she transferred from there to Stamford. When she left Stamford after about two months to begin a six-month stay at Otsego Manor, a conventional nursing home at the southern edge of Cooperstown, it was even easier for friends to come visit. Over that time, they could see sparks of the Lonni Sue they'd known so well. As her language returned, she began to greet them with a big smile when they showed up, as if to thank them for coming. She evidently had enough social intuition to realize that she was supposed to know these people, and therefore acted as though she did. "You might go for a minute having a conversation with her and think, My God, she's exactly the way she was before," Henry Weil recalled. "And then these drastic, dramatic lapses in memory would become apparent if you reminisced about some experience you'd shared or a conversation you'd had."

His wife, Rebecca, who had been even closer to Lonni Sue than Henry was, found these visits quite painful. "It was hard," she said. "She didn't know me, and, you know, probably for that reason I didn't come see her as often as I might have with someone else."

When Rebecca did stop by, she'd take Lonni Sue for a walk outside, maybe visit the little garden on the nursing-home grounds where vegetables and flowers grew. "She would just exclaim over each thing as if it was brand-new, and when I would come back, we would do it again, and it was all brand-new." It was wonderful to see Lonni Sue, and especially to see the joy and wonder Rebecca knew so well. "And yet," she said, "it's very hard to be with someone who doesn't know you, even if they're happy to see you."

Whether she knew them or not, Lonni Sue clearly enjoyed the visits. For their part, Maggi and Aline were not only pleased to have a sense of her life in Cooperstown, but also grateful at the help Lonni Sue's friends gave them in coping with the disruption of living out of their suitcases for weeks on end—telling them where to go for groceries, or what the best driving routes were when the roads were snow-covered, or helping them find places to stay. While Lonni Sue was at Otsego Manor, Chris Kjolhede, the pediatrician who had performed with her in the Fly Creek Philharmonic "orchestra," invited Maggi and Aline to stay in the guesthouse on his property.

During the months in Stamford and at Otsego Manor, as the bitter upstate winter gradually yielded to spring, her mother kept pushing Lonni Sue to draw. They hoped she would re-master this skill the way she had with the skill of feeding herself. Early on, Aline had had to wrap Lonni Sue's hand around a fork, guide it to pick up food from the plate, then direct it up to Lonni Sue's mouth. They would repeat this over and over, until, very gradually, over many weeks, she began to get it. "It was like programming a computer," Aline said. Each of the subroutines—hand on fork, fork to plate, food to mouth—had to be established, then hooked together into a sequence. "She started to get the sequence," Aline said. "She could go from A to B to C. But then she couldn't get back to A again." Lonni Sue didn't seem particularly frustrated by her limitations. She didn't slam the fork down in anger when she couldn't make her hand do what she wanted, the way you or I might have. She seemed to have no great desire to feed herself, and it evidently didn't occur to her that she was supposed to know how to do these things. "She was a happy camper, very amenable," Aline said. And she continued to look at the world around her with a sense of awe. "Your shirt, how inspiring!" Aline remembered her saying at one point.

It was the same with drawing as it was with eating. She wasn't especially interested in putting pencil or pen to paper for a long time. Maggi was relentless, however. She might be ninety years old, but the image of her mother Lonni Sue had described for Maggi's 2004 art show was still as valid as it had been when the girls were growing up: "on the threshold, with her hands on her hips, between being a mother and being an artist." Her daughter was disabled; her student had forgotten how to make art. In both roles, Maggi was absolutely determined to fix the problem as best she could.

Lonni Sue had drawn the bird outside her window in response to Aline's bird drawing. It was a copy, but it had just a bit of extra character that made it different, more essentially birdlike, than her sister's. So Maggi pushed her to be even a little more creative than that: she drew just a single, slightly curved line on white paper with a thin red marker. Then she handed Lonni Sue a blue marker, and told her to add her own. They took turns, each responding to what the other had just done, like musicians trading phrases in an improvisation. After a dozen or so turns, something began to emerge on the page: four stylized cats, sitting in a row. Maggi hadn't intended to draw cats when she began; Lonni Sue, who still had trouble initiating anything, certainly hadn't. There was no precise moment when the drawing changed from pure abstraction to something identifiable. At some point, both women just knew. Maggi had gently guided her daughter into making something from nothing—and the something was closer to Lonni Sue's whimsical, representational style than it was to Maggi's mostly abstract work.

The art lessons continued throughout the summer. Gradually, the programming began to take hold. The subroutines reestablished themselves in Lonni Sue's brain and nervous system. They hooked together in a sequence. And unlike the routines for eating or walking or getting dressed, they reactivated the neurons in Lonni Sue's brain where her creative spark lived—although neuroscientists still don't know where these brain cells might be located, or even what creativity is at the cellular level. By late summer, however, it was clear that Lonni Sue was recovering her art. "Finally," Aline said, "the little people came back." Tiny people sprinkled around the canvas had been a hallmark of Lonni Sue's illustrations for decades, ever since she'd taken that course on humor in illustration at the School of Visual Arts, in New York, early in her career.

Now, unprompted, the little people began to appear in her post-encephalitis drawings. "This was one of the first indications to us that those images inside her head were still there and she could convey them down her arm," Aline said. Before long, more of Lonni Sue's trademark images began to return as well—suns and moons, often with faces; horses; airplanes; faces hidden in the landscape. She began to do watercolor again. It still seemed that she couldn't recognize her closest friends. She didn't know Buzz, or Rebecca Weil, or Chris Kjolhede, whose guesthouse her mother and sister were now staying in. ("How do I know you again?" she asked him at least a dozen times when Maggi and Aline brought her over to his house for tea.) But somewhere in her ravaged brain, the technique of preparing and executing a watercolor was still present and accessible. Her brain knew what to do; her hands knew how to do it. For a long time, though, she needed a push from Aline or from Maggi before she'd do it. She could neither initiate drawings nor follow through with them on her own once she'd started.

. . .

During the six months or so when Lonni Sue was staying at Otsego Manor, Aline would sometimes spend days at a time up at the farm. Lonni Sue's personal and emotional relationships in Cooperstown had been ripped out and trashed by the encephalitis. Her financial ties were more stubborn. Aline took on the job of emptying out the farmhouse and studio, packing up and removing the artwork and Lonni Sue's other possessions, selling off the airplanes, disincorporating the dairy business, decertifying the landing strip with the FAA, and, finally, selling the farm. The whole thing had been a nightmare. Lonni Sue didn't have what some people call a "truck memo"—a list of assets and where to find them and what to do in case you're unexpectedly run over by a truck. Did she have a safe-deposit box? Aline didn't know, and Lonni Sue couldn't tell her. She had a bank account, presumably, and maybe several.

But what bank, and were there more than one? Who was Lonni Sue's accountant? What about the mortgage? "Lini had to sleuth, sleuth, sleuth," Maggi said, with empathy and admiration. Every time she found the right institution or the right person, she had to explain once again what had happened. Or almost every time. The woman who helped her out at Lonni Sue's primary bank, in Cooperstown, already knew. It was one of those old-style banks, Aline said, with a high ceiling and an enormous chandelier, and iron gates guarding the safe-deposit boxes. "This lovely lady," she said, "who was a stranger to me up until that time, sat across from me at a big, wooden desk." The banker was an artist herself, and she and Lonni Sue had struck up a friendship. "I was looking into her eyes to see my sister and what that friendship was," Aline said, "and she was looking into my eyes to find her friend in me." Some of the encounters involved that sort of deep connection. Others were a lot less meaningful, like the day when she had to spend several hours driving down to Sherman and back to close an account left over from the time Lonni Sue had lived there.

Aline had so much to deal with at the farm itself that she ended up spending many nights there—cleaning up, making repairs, going through records, and doing an exhaustive inventory of everything on the property. The process would end up stretching over more than two years. Buzz had moved into the house with a long-haul trucker named Amy whom he'd met and fallen in love with not long after Lonni Sue had gotten ill. About a year later, they were married. They couldn't very well begrudge Aline staying over, since her sister still owned the place. But sharing the small farmhouse wasn't always easy, Amy remembers. "Here we are in a brand-new marriage," she said, "and here's Lini. Buzz would be getting up to use the bathroom before going to work, and Lini would be in there showering at four o'clock in the morning."

According to Buzz, Lonni Sue had been somewhat relaxed about how he ran the farm. Aline wasn't relaxed at all. "She made me extremely mad, I can tell you," Buzz said. "There was this one time I put in a propane system at the farm to run the hot-water heater in the milk house," he said. Aline was concerned that he hadn't done the job right. "She was saying all these things, safety this, safety this. She was very compulsive about safety, Lini was. She got in my face about this thing, so I told her she could go shut off the gas herself if she wanted to." He remembers using a pretty severe expletive, although Aline insists he didn't. Now, looking back on it, Buzz and Amy are more understanding. "She's really very kind, and she had her sister's best interests at heart. She's just very . . . thorough," Amy said, after a meaningful pause. In the end, Aline's thoroughness turned out to be a good thing. An inspector ultimately agreed that the propane tank lacked a crucial safety valve. And because Aline was meticulous about listing every item on the property and whom it belonged to, Buzz was able to show that a rowboat on the property was his when new owners finally bought the farm and its remaining contents.

It's hard to imagine that Buzz and Amy didn't drive Aline a bit crazy, but her version of their time living together on the farm puts all the emphasis on the positive—as Aline's and Maggi's stories pretty much always do. "Buzz was very passionate about his work, and very articulate," Aline said. "He was under intense pressure to keep the dairy going, but he took the time to explain what was going on so I could do my role. It was so thought-provoking, his perspective."

In the end, she would spend a couple of dozen weeks at the

farm, a week at a time, before Watercolor Farm was all closed up and ready to be sold. Buzz and Amy weren't happy about moving. "I didn't know the farm was going to be sold," Buzz said. "When I found out, I even attempted to buy it, but it was just out of my league. And the Good Lord had other plans for me."

. . .

By late in the summer of 2008, Lonni Sue's language had fully returned. She could speak in complete sentences, and the sentences made sense. She had her nouns and verbs in all the right places. If you talked to her, she'd respond normally. If you didn't know she had profound amnesia, as Henry Weil and many others observed, you wouldn't realize anything was wrong, at least for a few minutes. It rarely took longer than that, however.

Like Henry Molaison, Lonni Sue couldn't form new memories, except with terrific difficulty, after hundreds of repetitions. She also had enormous trouble calling up not just specific episodes in her past but also the more general, semantic memories that H.M. had some access to.

The "Daddy died" story, in which Aline had to explain over and over that their father had passed away years earlier, was a good example: not only did Lonni Sue have no memory of this tragic event at first, but if you told her, she couldn't retain what for her was new information.

Another example was what Aline calls the "what happened" story. Lonni Sue knew that she had some problems with her memory, because her doctors and her family kept reminding her. She was completely unaware, however, of how profound her impairment was—naturally enough, because she had no way of grasping how much she'd once known. She didn't remember what it was like to have rich memories.

When the topic of memory loss came up, says Aline, "My sister would ask, 'What happened?' I'd answer, 'You got encephalitis.'"

"What's *that*?"

An inflammation of the brain.

"How *awful*. When did *that* happen?"

A few months ago.

And then, after a little while, Lonni Sue would ask, "What happened?" You got encephalitis. "What's *that*?" An inflammation of the brain. "How *awful*. When did *that* happen?"

They'd just spin around and around. It was just like the "Daddy died" story in that her effective memory span appeared to be less than five seconds, which would be dramatically worse than H.M.'s had been. In retrospect, however, Aline thinks that something else was going on. "Even while I was answering one question," she told me years later, "Lonni Sue would be inhaling to ask the next one. It was as though my answers weren't even registering." It seemed as if Lonni Sue's mind was so desperate to make sense of her situation that she couldn't absorb what she was hearing. She was really having a conversation with herself, oblivious to the fact that her questions were being answered.

But it turns out that she wasn't entirely oblivious. In a video Aline shot just a few weeks after Lonni Sue first became stuck in the "Daddy died" conversation loop, they talk about their father's death again, and the interaction is entirely different. At this point, it's still a surprise to Lonni Sue that their father passed nearly two decades earlier. She still sits bolt upright at hearing the news. The follow-up questions come less quickly, however, and more reflectively. Over many, many repetitions, the fact of Eddie's passing has somehow lodged itself in her brain, although she's not aware of it. She's not so much upset now as baffled. "I thought he was just staying with us . . . with my friends." She's baffled as well at the fact that she's having such trouble remembering more than the sketchiest of facts about her past. When Aline brings up Buzz and the farm, she seems to recall that she had a horse and a barn and cats (although all of these could be guesses), but there's little more than that. During the course of this conversation, she holds on a little better to the idea that her father is dead, but she's still surprised at several points when she hears, as though for the first time, how long ago it happened. When she finds out that it was in 1989, she calculates her own age at the time—she does remember the year she was born—and concludes that she was thirty-nine. She does this math at least three or four times.

Her language has fully and richly returned by now. When Maggi and Aline explain on the video that she hasn't been back to the farm for many months, and tell her about the encephalitis, she asks, "What did I do to open myself up to something so disastrous?" At another point, she asks, "Did my brain get eaten?" At still another she tells her mother and sister that "all the wonderful things in my memory got wounded." Her sense of humor has returned as well. At the beginning of the video, Aline and Lonni Sue are alone. Then Maggi walks in. Lonni Sue is wearing glasses; she greets her mother by jiggling them up and down, as though to say, "I see you." And then she laughs.

By the end of the summer of 2008, eight or nine months after the illness struck, the arguments in favor of keeping Lonni Sue in Cooperstown had become less compelling. Aline had tied up most of the loose ends of her sister's life, except for those involving the sale of the farm. She'd finished her sleuthing: the bank accounts were closed, the taxes paid. It's true that Lonni Sue's friends were there, but she didn't know them, and their visits had begun to taper off. So Maggi and Aline moved Lonni Sue one last time. Her rehabilitation therapy had ended when she left Stamford. She had long since become more or less medically stable. She could now walk and talk and feed herself. She could read. She could even play the viola. That had become clear a month or two earlier, when a friend brought her one and urged her to try. That language, like the language of art and the language of words, had not been wounded the way her memory had. But Lonni Sue couldn't live on her own: with such profound memory issues, there was no way she could navigate the world, either literally or metaphorically. So Maggi and Aline arranged for her to transfer to a facility down in New Jersey, where they could resume their own lives and help Lonni Sue to continue grappling with hers.

Lonni Sue was leaving the place where she'd finally found a sense of refuge and of peace, after years of searching. She wasn't troubled by the departure since she didn't remember anything about living there. Six months or so after the illness, and a few months before she left upstate New York forever, Lonni Sue visited the farm one last time. Amy was throwing a birthday party for Buzz. They invited the three Johnsons, along with a couple of neighbors. Everyone sat in the farmhouse kitchen where Lonni Sue had spent so many mornings, sometimes with Buzz, sometimes without, but always with her cats thirteen deep (they were no longer there; when it became clear that she wasn't coming back, Buzz had given them away). She could speak and respond to questions appropriately, but she didn't participate much in the conversation. Halfway through the meal, Aline told me, Lonni Sue looked over at Buzz and Amy. "Are you two married?" she asked. No, they said, they weren't. "You look as though you ought to be," she said. A few weeks later, they did get married. It was a lovely moment, Aline recalled. Whatever the virus had taken from Lonni Sue, it evidently hadn't robbed her of her intuition.

She didn't seem to recognize the kitchen, though. She didn't recognize Buzz. She might have retained some sense of familiarity about the farm. "As we were rounding the corner," Aline said, "before you could actually see the farm, Lonni Sue seemed to get excited." When they left a few hours later, she and Maggi tried to keep the memory alive. But by the time they reached the highway, Lonni Sue no longer remembered that they'd been anywhere special.

*Chapter 10*

# SECOND AND THIRD TURNING POINTS

It's November 2008. Lonni Sue Johnson has a headache. It's nowhere near as severe as the one she had nearly a year ago, as the encephalitis virus was escalating its assault on her brain. It's bad enough, though, and it's been going on more or less constantly for eleven months, ever since she was pulled back from the edge of death in Bassett Hospital. Lonni Sue's recovery has stabilized. She can walk now without assistance. She can talk in complete sentences, but her vocabulary is limited. When you talk to her, Aline says, it's like speaking English to someone who doesn't speak the language well. If you don't use the simplest words, you can't be sure she'll understand what you mean. She might act as though she does, though, answering questions in a generic sort of way. She appears to know, at least subconsciously, that she's *supposed* to know what you mean, and she doesn't understand that she has amnesia, so she plays along.

Maggi and Aline had hoped Lonni Sue's progress would have been greater by this point, but they understand that she might never have progressed at all. Still, they find it distressing that she seems to have no interest in initiating anything. She won't draw unless you urge her to. She seems happy enough, and she says things like "How inspiring!" whenever she's moved by something visually arresting, whether it's a painting or a sunset or a colorful scarf. She loves to eat, which worries Aline a bit. Lonni Sue was plump as a teenager, but slimmed down as she got older, not entirely by accident. She was so active when she lived on the farm that she couldn't have put on weight if she'd wanted to, but even so, she worried about every calorie. Now she sees no reason why an extra roll or two at dinner is a bad thing, and when Aline tells her "that's going to add pounds," Lonni Sue doesn't seem to care. She recognizes herself in the mirror, and she expresses no discomfort with how she looks. Aline and Maggi don't know what to do about that, and they haven't any confidence that she'll ever feel a sense of motivation or purpose about anything but eating her next meal when it's put in front of her.

Aline worries about her sister all the time. She was undoubtedly in the midst of worrying one morning that November as she took a walk, detouring around piles of raked leaves, when she ran into an old acquaintance named Amy Goldstein, who was returning from walking her daughter to Littlebrook Elementary School—the same school Aline and Lonni Sue had gone to as children. Amy had grown up mostly in Lawrenceville, the next town over, but she'd moved to Princeton in tenth grade. She was five years younger than Aline, so they didn't know each other at Princeton High. They met later, when they were in their twenties, at international folk dancing. It's arguably the nerdiest of all known forms of recreation. Princeton's nerd population is so high that this relatively small town had two rival international dance groups, each run by its own charismatic leader. Most people were either Tuesday night dancers, with

Leo, or Friday night dancers, with Jerry; only a few brave souls went to both. Aline was one of those, but she got to know Amy at Leo's group.

They hadn't seen each other since the late 1970s, however—Amy had left Princeton for New York after that, and had only returned a few years ago—so they caught each other up. Amy, it turned out, had become a professional puzzle writer. She specialized in word-search puzzles, the kind where you look for words embedded in a grid of letters that appear at first to be gibberish. "People who are really into crosswords and other word puzzles look down their noses at word searchers," she said, "but I happen to be good at making them. I sell books. It's very nice."

When it was Aline's turn, she began just as she had with me. "Have you heard what happened to my sister?" she asked. Amy hadn't heard, but she knew very well who Lonni Sue was. Her *New Yorker* cover with a line of people carrying presents forming a Christmas-tree shape as they wait at the post office was one of her all-time favorites. She knew who had drawn it, and she knew more generally who Lonni Sue was professionally. Amy hadn't realized, however, that Lonni Sue was related to Aline. "So we're walking and chatting," Amy said, "and she's telling me this awful story about Lonni Sue, and I'm like, 'Oh my God, this is terrible.'"

Amy wanted to help in any way she could. She wondered if Lonni Sue might be interested in trying some word-search puzzles. Aline didn't really know. Lonni Sue had loved words and wordplay before the illness, so the idea wasn't crazy. She had also loved to draw, though, and that didn't seem to give her a lot of pleasure anymore. Nevertheless, it would be rude to refuse the offer, so the women walked over to Amy's house, where she gave Aline three puzzle books. "I thought, 'Oh, my gracious, what if Lonni Sue isn't interested?'" Aline said later. "'Amy's been so generous.' But then I thought, 'Well, okay. We'll give it a try.' We try to give everybody's ideas a chance because we certainly have no clue of what will help my sister." Amy herself wasn't counting on it doing any good, not even slightly. "I gave her the stuff," she said, "and Aline thanked me, and that was pretty much our conversation. I told my husband about this and more or less forgot all about it."

About a year later, Amy ran into Aline, again by chance—she was out with her husband and her son this time—and Aline told her the astonishing story of what had happened next. Aline gave her sister the puzzle books, not expecting much. To her surprise and Maggi's, Lonni Sue didn't just enjoy the puzzles; she loved them. In retrospect, maybe it shouldn't have been surprising. She couldn't have done crossword puzzles, which usually call on the puzzle-solver to delve into her memory to find a word to match its definition. With word searches, however, memory isn't required. The words are already there, right in front of you. All you need to do is find them and circle them, which "remembers" for you that you don't need to find that particular word again. If you get interrupted, you can come back a minute or an hour or a day later and pick up where you left off: the puzzle itself tells you what words you've already found. Not long after she started working the puzzles, Aline said, "Lonni Sue fluttered her hands next to her ears and said, with real excitement, 'This helps me clear my mind!'"

Maggi and Aline were thrilled to see Lonni Sue so enthusiastic. Nobody had to urge her to work on the word searches. It was the first time since before the illness that she could work on her own at her desk, concentrating, unprompted, for hours on end—or even for minutes on end. She could play a short piece on the viola on her own as well, but it wasn't the same. She would follow the sequence of notes on a page carefully, one by one, from beginning to end, but if she got distracted, she wouldn't know where to resume. She wouldn't choose to pick up the instrument on her own, and she would never play for long.

This was completely different. After a couple of weeks, Lonni Sue had worked her way through all three books. When she was finished, she turned to her mother and her sister and asked, with a real sense of urgency and alarm, "What am I going to do now?

Should I erase the ones that I've done and start again? Or . . . could you please get me some more books?" Maggi said she would. She had a few other things to do first, though, so in the end she didn't get to the bookstore until the next day.

Lonni Sue couldn't wait that long. The puzzles had given her a purpose for the first time in many months. Ever since she'd been a teenager, she'd been relentless about working toward success, first as a musician and artist, then as a pilot. She'd worked constantly, almost compulsively. Now, for nearly a year, she'd been almost completely inert. She had a pretty good excuse, given that a virus had invaded her brain and nearly killed her, and she showed no impatience or discontent with the situation. She was perfectly cheerful most of the time. Perhaps this was because she'd forgotten what it felt like to throw herself wholeheartedly into something. If that was the case, she might never care about anything enough to work at it again. But suddenly, here she was, desperate to work, and she had nothing in front of her to work on.

So she began to create her own puzzles. She'd figured out more or less right away that the words in many word-search puzzles aren't random. Each one conforms to a theme, so her puzzles began to have themes as well. At first, these focused on obvious, concrete categories—the names of colors, for example, or of shapes. Then she started to get more fanciful, with categories like "What are your favorite things in life from A to Z?" Maggi and Aline couldn't believe it. It was gratifying enough to see Lonni Sue excited and motivated, but now she was being creative as well. They were thrilled, but they were also concerned. Where was the art? This was a person who lived and breathed art. Since she'd sketched the bird in a tree back at the rehab in Stamford, her drawings had become increasingly colorful and complex. But she wouldn't draw without prompting. Now she had her creative energy back, and she was making . . . grids of letters.

Then, on the thirteenth puzzle, a small pear and a small apple appeared on the page. From then on, the illustrations became more and more complex and exuberant. One especially vivid example has a twenty-by-twenty array of letters (approximately, since some spots on the array are blank). The words embedded in the grid are also printed carefully above and below it, suspended in cloud-shaped balloons like the ones comic-book artists use to show that a character is thinking rather than speaking. Each balloon contains words that start with the same letter—"mist, movement, music, magic . . ." or "enthusiastic, eyebrow, ears . . ." Smack in the middle of the grid is a brilliantly colored drawing of a woman wearing a pleated skirt ("pleats" is one of the words in the grid). She's also wearing an apron decorated with the sun, the moon, several stars, and dozens of musical notes. Her legs suggest she's running or leaping, her curly brown hair and the apron strings and a scarf flying out behind her. She's reaching up with both braceleted hands toward a collection of objects floating above her head—another sun and moon and stars, but also a kite and a cat and a pear and a seashell and a baseball, and a few teacups.

To this day, Amy Goldstein has seen only a fraction of the puzzles Lonni Sue created. Of these, her favorite has a knitting theme. It came about as a result of Aline's relentless effort to probe her sister's memories and abilities and push her to recover as much as possible. "I was blown over when she first tied her shoe," Aline said, so naturally she wanted to see what else her sister might be able to do. Lonni Sue had knitted in high school. Could she still? To motivate her to try, Aline suggested she knit a scarf for their mother. Lonni Sue wasn't especially interested, so Aline did two rows and coaxed Lonni Sue into doing one. She did, grudgingly. They alternated that way for a couple of days. Then Lonni Sue had an insight. Nobody had dared hope she'd ever have an insight again. "Knitting is like puzzles," she said. Aline asked her how. "Because of the rows and the columns," she said. She hurried to her desk, and began to draw.

The finished puzzle has a drawing of two knitting needles at the top. A scarf unfurls from the needles; each stitch is a box with a letter inside of it, spelling out words like "yarn" and "wool" and "sweater" and other knitting-related terms. "This is brilliant," Amy said. "It's absolutely brilliant. Honestly, I would have been really proud to have come up with this idea myself." Another puzzle has a word grid embedded in a wire coat hanger, with the words inside the grid all relating to clothing. Ever since Aline showed Amy some of the puzzles, she has wished she could take all of Lonni Sue's work from that period and look for more gems that Aline and Maggi might not realize are special. "These aren't just art," Amy said. "They're art-plus. She's bringing everything she has to these."

Once Lonni Sue began creating her own puzzles, there was no need to buy her more word-search books. Instead, Maggi bought her a dictionary. She browsed it endlessly for inspiration. She carried it with her everywhere she went in her circumscribed orbit—in the facility where she lived, on visits to Maggi's house, while waiting for a physician's or a dentist's or an eye doctor's appointment—along with paper and pens and pencils so that she could work on her puzzles during every spare moment. Within a year, the dictionary's crisp new pages were worn and soft-edged. The binding was broken, and had been taped and re-taped. At about the same time that she began creating her puzzles, the headache that had tormented Lonni Sue for nearly a year vanished. It has never returned. Perhaps it had been an outward symptom of her unconscious impatience to get back to some sort of productive work. Or perhaps it was the lifting of the headache that allowed her to concentrate on the puzzles. Doctors had never been able to pinpoint the cause, so either explanation is possible.

When she started out, the puzzles were pure word searches. Then they became word searches decorated with images, which were increasingly important. "They flowered," Aline said. "There was more and more complexity in her relationship between word, image, and grid." Then, about four and a half months after the puzzle-making began, things veered off in an unexpected direction. Lonni Sue developed *hypergraphia,* which is an overwhelming desire to write. Her puzzles had always included lists of the words hidden within the grid, organized alphabetically. Now she was interested in the lists themselves. "She just plastered the words over the pictures," Aline said. "They squeezed out the drawings, and it was just about words, words, words, words."

Before long, Lonni Sue had nearly given up on pictures. She began listing the letters of the alphabet vertically along the left-hand side of the page, then filling out the rest of the page with words beginning with each letter. Hypergraphia, which may have afflicted Fyodor Dostoyevsky, Lewis Carroll, Vincent van Gogh, and Robert Burns, is often associated with epilepsy centered on the hippocampus and the medial temporal lobe. Lonni Sue would write long into the night. It became all-consuming. She wouldn't sleep. Maggi and Aline began to fear for her health, but Lonni Sue was elated. She began creating "alphabet sentences," each one exactly twenty-six words long, in which the first word began with an "a," the second with a "b," and so on through "z." She savored words like you might savor morsels of chocolate, rolling them around in her mouth. She came across the word "alluring," and insisted that Lini write it down for her so she could remember it. "She would rediscover 'alluring' over and over again, for two weeks in a row," Aline said, "then would fall in love with another word." At one point, Aline asked her sister how she thought the brain is organized. "I think mine is organized on the alphabet," Lonni Sue replied.

It took about a year and a half, but the hypergraphic stage slowly passed. Looking back on it, Aline thinks that the word lists and the illustrated puzzles served as a sort of jungle gym Lonni Sue's brain had devised to exercise itself. It gave her a set of mental workouts to rebuild her shattered mind, drawing on the love of words and her skills as an artist that had been so important to her. Her memory was mostly gone, but the relentless drive to work on something creative remained. She just needed a way to channel it.

As the hypergraphia lifted, Lonni Sue's puzzles changed dramatically. They were different from anything she—or anyone else—had done before, and the design hasn't changed since late in 2010. Each puzzle, usually executed with brightly colored pens and markers, now consists of four pages, Aline explained. The first, which she calls the cover page, starts with a border of squares or rectangles that form a frame. She writes letters of the alphabet inside the squares. Sometimes she'll do little drawings of objects or creatures whose names start with the appropriate letter—an apple for "A," a bumblebee for "B," and so on. If the number of squares isn't an exact multiple of twenty-six, she'll fill in the blanks with stars or moons or big, colored dots. Inside the frame she puts a drawing that represents the puzzle's theme. It might be a category, like "flowers" or "things to eat," or it might be an activity, like "swimming."

The second page always has what Lonni Sue calls a "story" at the top, but it's really no more than two or three sentences that include some wordplay or maybe some sort of philosophical observation about art or music or life. These aren't especially sophisticated—in one, for example, she declares, "Art is a way of expressing feelings we have"—but the fact that her damaged brain is capable of having such uplifting thoughts is deeply gratifying to Maggi and Aline. There's also an alphabet "sentence"—an "A" word followed by a "B" word, and so on. After that come twenty-six more sentences, each constructed of words that all begin with the same letter. This is pretty clearly a preserved remnant of her hypergraphia, Aline says. On page four she uses the words of the "story" to make a word-search letter grid, with at least part of the story repeated in a box at the center of the grid. And on page three, her plan is always to blow up that box and replace the actual letters with drawings of people in the shapes of letters. She never carries this out, however.

In five years, in fact, she's never completed a single one of these four-page puzzles. She always gets distracted by new ideas, and can never pick up the thread of what she'd originally been thinking. Nevertheless, and in spite of their repetitive format, the puzzles are legitimate, if unfinished, works of art, with an internal logic that gives her mind something to hold on to. The puzzles engage and reassure her. She'd rather be immersed in what Aline calls her "puzzle world" than doing anything else. "I work on my puzzles when I'm allowed to," she'll sometimes say, plaintively, to anyone who will listen. Pretty much the only time she isn't allowed is when she's sleeping, or getting some exercise in a swimming pool. She accepts all of these impositions on her time, but she doesn't always do so happily.

None of this would have happened if Aline hadn't run into Amy Goldstein. Of course, Lonni Sue's brain might have found another way of organizing itself anyway. The drive and focus that had made her such a successful illustrator were evidently left intact when the encephalitis destroyed her hippocampus. If she hadn't latched onto word-search puzzles, she still might have found a purpose in life. Maybe not, though. The encounter was purely serendipitous, and the fact that Amy had turned out to be a puzzle writer made it doubly so. How many puzzle writers are there in the world, anyway? It was a complete surprise to the Johnsons, moreover, that Lonni Sue responded so powerfully to the word searches. She'd loved puns and wordplay before the encephalitis, but not this particular form of wordplay. No one could have predicted any of it.

Here's another thing nobody could have predicted. In December 2009, a little more than a year after Lonni Sue began her love affair with word puzzles, Aline was in downtown Princeton to do some errands. Lonni Sue was still in the grip of hypergraphia; the compulsion to write hadn't eased yet. Aline and Maggi didn't know whether it ever would. Aline was exhausted. She was still driving up to the farm every other week to work on packing up the house and putting her sister's affairs in order. She hardly knew which end was up. "I was in a daze most of the time," she said later. Aline was walking up Nassau, Princeton's main street, which separates the town where she grew up from the university where she got her degree and where she worked for many years. As she trudged along the sidewalk, she passed Landau's woolens shop and thought about Lonni Sue's friendship with Robert Landau, which stretched back more than thirty years at this point. Tired as she was, she had made a vow that she would be in contact with every one of Lonni Sue's friends and colleagues—not just to inform them, but in case they might help her and her mother figure out a way to turn the tragedy into something positive. "I should tell Robert what happened," she realized, so she stepped inside.

Landau didn't really know Aline. They'd never been formally introduced. He did know Maggi, and he also knew that every once in a while a younger woman who resembled Lonni Sue would come into the store. It hadn't been hard to figure out who Aline must be. "I would say, 'How's your mother?'" he recalled, "and she'd say, 'Fine.'" Then, "How's your sister?," and the reply was the same. That was pretty much the extent of their interactions. This time, Robert began the old routine. Maggi was still fine, he learned, but when he asked about Lonni Sue, he was startled at the answer. "She's had this traumatic incident," Aline said. They sat down, and she told him what had happened. "I hadn't heard anything," Robert said. "If Aline hadn't come into the shop, I'm not sure how long it would have taken me to find out."

Aline explained the situation in detail. Lots and lots of detail, including plenty about the medial temporal lobe and anterograde amnesia and more. At several points during the conversation, Robert admitted, "I had no clue what the heck she was talking about." He's a smart and educated guy, but he's a small businessman, not a neuroscientist.

His wife, however, is a cognitive scientist, which is pretty close. Like Lonni Sue, Barbara Landau had grown up in Princeton. They were both alumnae of Princeton High School, with Barbara graduating a few years before Lonni Sue. Her father had worked at RCA, just as Lonni Sue's father had, although not directly with Eddie Johnson.

Barbara Landau had had a distinguished career in psychology, linguistics, and neuroscience, including jobs at Columbia University, the University of California, Irvine, and now, Johns Hopkins. "My wife is really interested in the brain," Robert told Aline during their conversation in the store. "Maybe you'd like to talk to her about Lonni Sue." Aline didn't strike him as being very enthusiastic at first, but Robert hadn't made it clear that Barbara was *professionally* interested in the brain. Aline was imagining someone who liked to read books by Oliver Sacks, and who maybe audited a neuroscience class or two at Princeton. "I pictured her as one of those gray-haired people from the community who sit in the back rows," she said later.

Once she realized her mistake, however, her attitude flipped. Right from the beginning—Aline thinks it was probably within the first ten days after they got that first phone call from Bassett Hospital, when the enormity of what had happened was still sinking in—she and Maggi had promised each other that they'd try to find a way to turn the tragedy of Lonni Sue's illness into something that could help people understand the brain better. They hadn't figured out a way to do that yet. They'd been entirely focused on getting Lonni Sue to walk, talk, and feed herself. Then they'd pushed her to begin drawing again. When Amy Goldstein's puzzles brought her to life, they were surprised and pleased. Lonni Sue could have ended up much worse off. She had recovered many of her faculties, and she'd found a purpose in life, albeit a very narrow one. She was almost invariably warm and cheerful and happy. Not everyone with such severe brain damage is fun to be around. Lonni Sue was.

Still, while her mother and her sister hoped she would continue to improve, Lonni Sue's situation remained pretty much a disaster. There was clearly no prospect of her ever returning to anything like normal. When Robert Landau explained that his wife studied the brain, Aline thought again of the promise she and her mother had made to each other. Aline and Maggi were thinking about advancing science itself, which had been so important to Ed Johnson. But there's a larger context as well. If Lonni Sue's amnesia could help them understand more fully how the brain works, then this terrible misfortune could contribute in at least a small way to helping people with memory loss. That wouldn't just benefit the thousands of patients who survive HSV encephalitis or brain injury.

It could also potentially help the tens of millions who suffer with Alzheimer's and other forms of dementia, which also affect memory but in different ways. (Alzheimer's, for example, attacks not only the medial temporal lobes but also the frontal lobes, which affects victims' language, judgment, and cognitive functions.) Looking at memory loss from many perspectives, through the lens of many different memory disorders, might lead to a treatment—maybe not right away, and only as part of a much larger scientific enterprise. But every bit of progress can help.

Aline knew about H.M. from her course work, so she understood that while it happens only rarely, a single patient can sometimes revolutionize an entire field of research. When Henry Molaison died, in December 2008, Suzanne Corkin had declared at his memorial service that "we say good-bye to him with respect and with gratitude for the way in which he changed the world and us. His tragedy became a gift to humanity." If Lonni Sue could do a tenth as much for neuroscience as Henry Molaison had, her tragedy could be a gift as well. Maggi and Aline had also decided that if neuroscientists were interested in studying Lonni Sue, or if journalists wanted to write about her, they'd be allowed to use her full name. H.M. had been known only by his initials until after he died, but they felt that Lonni Sue's extraordinary life before her amnesia was an important part of the story. They wanted the public to understand what had been lost. So yes, Aline would definitely talk with Barbara Landau, and the sooner the better.

When Robert broached the idea to Barbara, it was his wife's turn to be skeptical. "Are you kidding?" she asked. "I don't know anything about amnesia, and I don't work with adults." Her specialty was development—trying to understand how kids' brains and cognition grow and mature. "Well, couldn't you just talk to her?" Robert asked. "I think she just wants to talk." Okay, okay, Barbara said. It couldn't hurt just to talk.

When Barbara made the call, she fully expected to listen patiently before explaining to Aline why this wasn't really of professional interest to her. That's not how it happened. They ended up having a three-hour conversation. Aline told her the whole story, from the phone call to the months of recovery to the return of Lonni Sue's art, little people and all. She told Barbara about the puzzles and the dramatic change it had made in her sister's mental state. As she talked, Barbara's skepticism began to weaken. "Actually," she thought to herself, "this is unbelievably fascinating." She didn't know much about amnesia, and she didn't know anything about art, but how could this not be really interesting?

This wasn't an entirely rhetorical question. Neuroscientists work on many different aspects of the brain. Experts on memory don't necessarily know much about perception, or consciousness, or depression, or Parkinson's disease, and vice versa. For all Barbara knew, her colleagues who actually did know something about memory and amnesia might see right away why the case wasn't especially interesting after all. Barbara's personal fascination with Lonni Sue could easily be influenced by the fact that Lonni Sue's artwork was hanging all over the Landaus' house. When you're doing science, the Nobel Prize–winning physicist Richard Feynman once said, "The first principle is that you must not fool yourself, and you are the easiest person to fool." Not wanting to fool herself as she ventured outside her scientific comfort zone, Barbara consulted with some colleagues who did know something about the topic.

This is really, really interesting, they told her.

So Barbara stopped by to visit one day the following spring. She ran Lonni Sue through a couple of basic memory tests her colleagues had helped her put together. "I took a bunch of notes," she said. She went back to Baltimore, talked to some more colleagues, then returned with another Hopkins neuroscientist named Mike McCloskey that summer. "We did another visit," she said, "and began to really poke around scientifically. That was the beginning of it."

It wasn't simply a matter of putting Lonni Sue through some sort of standard regimen. There isn't really a standard. When Brenda Milner and Suzanne Corkin began testing Henry Molaison in the late 1950s and early 1960s, they gave him some very basic memory tests at first. Then, as they grew to understand the outlines of his memory loss, they modified the tests and designed new ones. A generic test can tell you only so much. In typical scientific research, McCloskey explained to me when I visited Johns Hopkins, you tend to know what you're looking for. Then you map out a series of experiments to try to find it.

With an individual patient like H.M. or Lonni Sue, it's different. The patient is the experiment. She presents the scientists with a set of symptoms, or characteristics, and it's your job as a scientist to figure out what those symptoms are telling you. "You do preliminary testing to try to probe around and see what's working well, what's not working well," he said, in his gentle, quiet voice, which is a bit reminiscent of Mr. Rogers. After you've got the general lay of the land, you can focus on aspects of memory loss that are particular to that patient. "It tends to be kind of opportunistic," he said.

In Lonni Sue's case, the opportunities seemed especially intriguing for a couple of reasons. One was the extreme depth of her memory loss—the density of her amnesia. She was like Henry Molaison in that she couldn't form new memories, and that her episodic memories of the past were almost entirely gone. Unlike H.M., however, she also had relatively few semantic memories about her life. She knew she'd been a pilot and that she'd lived on a farm, but she didn't know that she'd been married for ten years. She couldn't recognize photos of some of her relatives. There were a few bizarre exceptions. She still knew the voice of her childhood friend Emily Speagle on the telephone, even though Speagle's voice had presumably changed somewhat over all those decades, and even though the women had been in touch only intermittently since high school. But most of it was gone.

A more important difference between Lonni Sue and most other patients with medial-temporal-lobe amnesia was the breadth of her expertise before the encephalitis struck. She'd been a serious amateur musician, a pilot, an extraordinarily successful artist. She was accomplished in all sorts of fields that required extensive learning and memory, and which depended on an enormous storehouse of knowledge. Exploring what she'd retained and what she'd lost in each of these areas would be an unprecedented opportunity for the scientists, and for the field of memory research in general. H.M. had made it possible for Milner, Corkin, and their collaborators to answer many basic questions about what memory is and how it works. Now Lonni Sue promised to answer a whole new set of deeper, more complex questions.

She was unlike any amnesia patient who'd been studied up until that point. H.M.'s highest professional achievement had been winding copper wire around electric motors. E.P., an amnesia patient studied by the neuroscientist Larry Squire, was an engineer, but Squire didn't explore his professional knowledge extensively. The only amnesic in the literature with a comparably rich and complex life before his hippocampus and nearby tissues were destroyed was Clive Wearing, a Cambridge-educated British musician. Wearing was a professional composer, organist, conductor, and vocalist. He was an expert on the music of the Renaissance—specifically on the music of the Dutch composer Orlando de Lassus—and a music producer for the BBC's Radio 3 network. In March 1985, he came down with HSV1 encephalitis, the same infection that attacked Lonni Sue. Wearing worked long, punishing hours, just as Lonni Sue did. As a result, his wife, Deborah, wrote in her 2005 memoir, *Forever Today,* "It was hard to pinpoint exactly when exhaustion turned into the first stirrings of illness. Sifting through the weeks and months before, I cannot know for sure."

Like Lonni Sue, Wearing's first symptom was a bad headache, followed by a fever and increasing confusion. Lonni Sue's condition worsened without anyone really noticing, but Deborah could see her husband getting progressively sicker over the course of several weeks. His headache intensified. His fever kept pushing higher, and he became increasingly confused. Maggi and Aline have always wondered if they could have prevented some of Lonni Sue's brain damage if they'd been up at Cooperstown for the holidays. They probably would have noticed her worsening condition. They might have gotten her to the hospital sooner. But Deborah Wearing did get her husband to see a doctor early on—two different doctors, in fact. Both waved her away, saying it was just a bad flu. Deborah Sentochnik, the infectious-disease specialist at Bassett Hospital, told me that HSV encephalitis is always on doctors' minds, despite its rarity. She might have overstated that case.

In the end, Clive Wearing survived his encephalitis in worse shape than Lonni Sue. He retained his speech and vocabulary. He could still read music, play the organ, and conduct a choir. He recognized Deborah and recognized his children. Nevertheless, as the late Oliver Sacks wrote in a profile for *The New Yorker* in 2007, "Clive showed a desperate aloneness, fear, and bewilderment. He was acutely, continually, agonizingly conscious that something bizarre, something awful, was the matter." Wearing was typical in that he couldn't remember anything that had happened to him a few minutes earlier, but unlike both Henry Molaison and Lonni Sue, he found this terrifying. "It was as if every waking moment was the first waking moment," Deborah Wearing wrote. It made him frantic. "I haven't heard anything, seen anything, touched anything, smelled anything," he would say. "It's like being dead."

Clive kept a diary of sorts, but the entries were all more or less the same. They're kind of appalling to read. This is just part of what he wrote on March 25, 1985:

> This (officially) confirms that I awoke at 9:05 am this morning
>
> I woke again at 2:06 pm (this time properly) . . .
>
> 5:00 I finally awoke
>
> At 9:40 pm I awoke for the 1st time, despite my previous claims
>
> I was fully conchase [sic] at 10:35 pm, and awake for the first time for many, many weeks.
>
> Proper consciousness at 1:19 am

It went on like this, page after page, week after week. Lonni Sue wrote obsessively, too, during her hypergraphic period, but it was all word lists, not this sort of endless recurring loop. Wearing's constant sense that the world had just come into existence wasn't limited to his own consciousness. At one point, Deborah arrived at the hospital to find her husband holding a piece of chocolate in his left palm. He covered it with his right hand, then uncovered it, as if, she writes, "he were a magician practicing a disappearing trick."

> "Look!" he said. "It's new!" He couldn't take his eyes off it.
>
> "It's the same chocolate," I said gently.
>
> "No . . . look! It's changed. It wasn't like that before . . ." He covered and uncovered the chocolate every couple of seconds, lifting and looking.
>
> "Look! It's different again! How do they do it?"

He was so distressed and agitated so much of the time that there was no way you could sit him down for extended testing. Henry Molaison had been an amiable, cooperative subject who didn't have any special talents or reservoirs of knowledge to probe. Clive Wearing had plenty of both, but wasn't capable of working with scientists. Lonni Sue was accomplished in a variety of unrelated fields, and she was also, and continues to be, a nearly perfect test subject. "She's incredibly cooperative," Barbara Landau said. She sometimes gets impatient when a particular test goes on too long. She'd rather be working on her puzzles than taking part in experiments, and she's always got her tote bag filled with half-finished puzzles to turn to if she's given the chance. But it's usually easy to coax her into continuing. "Just a little bit longer" is generally all it takes, along with another reminder that while the tests themselves might not make sense to her, the research will help other people.

Lonni Sue is even comfortable inside an fMRI scanner—that is, a functional magnetic resonance imager. An fMRI measures changes in blood flow in the brain, which correlates with brain activity. The scientists were worried that she'd feel claustrophobic inside the machine, as many people are. It wasn't a problem. They also worried that she might drift off to sleep inside the scanner, then wake up and panic because she wouldn't know where she was. "At first," McCloskey said, "we even had a message that was up there all the time at the bottom of the viewing screen that said something like, 'Hi, Lonni Sue, we're scanning your brain. Please keep still,' or something like that." It was never an issue. "It's not easy to lie in one of those scanners for a couple of hours," McCloskey said. "Have you done it?" About a year after the testing on Lonni Sue had begun, Emma Gregory, a postdoctoral fellow at Hopkins who was working with McCloskey and Landau, presented some preliminary research at a neuroscience conference. Other scientists kept coming up to her and saying how envious they were that she got to work with such a ridiculously cooperative subject. "They were almost salivating," Gregory said.

The real problem with Lonni Sue was that she presented an embarrassment of riches. You could explore just about any question a neuroscientist could ask about memory or cognition with her, Mike McCloskey said. "It's almost that there's too much that you could do," he said. When we spoke, in 2014, after they'd been studying her for three years, McCloskey felt that they'd barely scratched the surface. That's the glass-half-empty perspective. The glass-half-full version is summed up nicely by Neal Cohen, Suzanne Corkin's former graduate student and Henry Molaison's onetime driver. Cohen hasn't worked with Lonni Sue himself, but he knows about her and has been following the case closely. "I think she might be the most interesting amnesic to have been studied in this level of detail," he said.

The most obvious place to start with Lonni Sue was with her art. It was not only the activity that had defined her life over the previous four decades or so; it was also the aspect of Lonni Sue that Barbara Landau knew best, based on their long-standing relationship. Landau was especially intrigued by the fact that Lonni Sue's impulse to make art had returned, despite the terrible damage to her brain, and that it showed at least some elements of her pre-encephalitis work. At about the same time that she, Mike McCloskey, and Emma Gregory had begun testing Lonni Sue in earnest, Barbara went to an exhibition at the Walters Art Museum in Baltimore titled Beauty and the Brain. It was coproduced by the Walters and by the Johns Hopkins Brain Sciences Institute; the idea was to show people works of art, survey them to ask what did or didn't make the artworks appealing, and finally put some of the respondents into an fMRI to see which parts of their brains became active while viewing pieces they liked and didn't like.

Barbara cornered the museum director, Gary Vikan, and told him she had an amazing idea for another exhibition: the museum could display Lonni Sue's art, both pre- and post-encephalitis, along with scientific explanations of what had happened to her. Vikan loved the idea. A year later, in September 2011, with enormous help from Aline and from Barbara Landau, Puzzles of the Brain: An Artist's Journey Through Amnesia opened and ran for two months. In January 2012 it moved to the Morven Museum in Princeton, where it ran for another five. It's now packed up and sitting in the basement of the Johnson house in Princeton. Aline is determined to show it again somewhere, perhaps in Cooperstown.

During the decades when Suzanne Corkin and her colleagues studied Henry Molaison's amnesia, it was easier to bring Henry to the scientists in Cambridge every so often than to bring an entire lab's worth of researchers to Henry's nursing home in Connecticut. The Johns Hopkins team was much smaller, and the Landaus still had a home in Princeton, so the scientists traveled up to New Jersey once a month or so to administer memory tests in the Johnsons' dining room, at the table where Lonni Sue and Aline had sat as children. Those long-ago chamber-music evenings had taken place in the adjacent living room. Lonni Sue's favorite vision of her mother from childhood—the one she'd written about, of Maggi in her apron in the doorway between the kitchen and the studio "on the threshold, with her hands on her hips, between being a mother and being an artist"—had been set just a few feet away from where the testing was done. Lonni Sue showed little evidence that she remembered any of that now.

Studying Lonni Sue's artistic talent and its relationship to memory and amnesia proved to be more difficult than the scientists had anticipated. "It turns out to be a complicated set of questions," McCloskey said. She was wildly creative, but neuroscientists are far from understanding the neurological basis of creativity. They don't know what it is, exactly. They don't know how to measure it. They don't know where it lives in the brain. Beyond that, the creative spark that Lonni Sue displayed in her pre-encephalitis art was only part of what made the work compelling; the intelligence behind the work, as the designer Tom Hughes had written, was also essential.

Mike McCloskey likes the analogy of creative writing, which might be a little easier for nonartists to think about. At the most basic level, he said, you have to have a rich vocabulary. Then you have to be able to use that vocabulary to compose compelling sentences, and weave those sentences into paragraphs. You also need to understand how to string the paragraphs into a narrative. And you need an idea of what the story is that you want to tell. It's a mix of basic skills plus knowledge plus the eye (or ear) of an artist. "Our understanding of how the brain works might allow us to attack some of those lower-level skills," McCloskey said, "but how you conceive of the novel and how you get from your conception to the words on the page is the sort of thing neuroscience is only starting to address."

So the Johns Hopkins team began with those lower-level skills and knowledge: a straightforward assessment of Lonni Sue's semantic memories about art—information she had taken in during her education and career, and that might or might not still be accessible. They showed her seventy images of artworks in succession. The scientists asked her in each case if she could name the painting or the artist. She identified only the *Mona Lisa* and *The Last Supper,* which is pretty shocking for someone who studied art history in college and became an artist herself. Or, to be more accurate, those were the only ones she recognized among the works that people of a similar age and education, but without brain damage, would probably be able to identify. Ten of the artworks had been drawn by Lonni Sue herself, before the encephalitis. She recognized all of them as hers. A few others were done in a style broadly similar to her own. "I might have done that one," Lonni Sue said of these. "One of the things we've kicked around doing," McCloskey told me when I visited Johns Hopkins in 2014, "is to take some of her artworks and do a bit of Photoshopping on them, and see if she can distinguish the ones that we've manipulated from ones that we didn't."

That test showed that Lonni Sue doesn't simply recognize her own art: she recognizes style, which is more elusive. "Whatever it is that allows her to recognize her own style," Landau said, "might be incredibly complex." It's not episodic memory, obviously, since it's not based on specific incidents. It's a more general kind of knowledge—but it's not clear that it's semantic memory, either. Or perhaps it is, for the subset of people who work in art. A critic can put into words what makes a Monet broadly different from a Manet, even if you or I couldn't. "I don't think we know how to characterize that sort of memory," Landau said.

Or it could be a lot simpler than that. Maybe it's not the style of her work overall. Maybe she picks up on just the simplest characteristics of her art—the suns and moons and stars and little people. "The way that I think about it now," Landau said after a couple of years of working with Lonni Sue, "is that when she did the Princeton Poster, when she did pieces for *The New Yorker,* she was drawing on a huge amount of stored knowledge about the world." She was a history buff. She had this huge reservoir of information that she could bring to bear on the creative process. That, Landau suspected, was what you saw in all of the work that she did, these incredibly intelligent, rich pieces that were loaded with detail—not only facts, but facts that she spun and elaborated on in unusual ways. "I think the art now has none of that fact-based knowledge, because it just isn't there," Landau said, "but it still retains this whimsical creative inventive spin on things. I don't think we've been able to quantify that, but that, to me, is one of the biggest mysteries and puzzles here. I find it fascinating. I think we haven't really gotten a handle on it."

The second set of art-related tests involved quizzing Lonni Sue on the mechanics of producing art with various media—watercolor, acrylic, oil, pen and ink—and on the principles of design. Relying heavily on Maggi's expertise, they developed a set of questions that asked things like: What are the steps you have to do to produce a watercolor? What is a wash? ("Don't ask me what it is!" Emma Gregory said to me as she was describing the experiment.) How do you draw lines to create the illusion of three dimensions? Things like that. Unlike her performance on the "What is this work of art and who created it?" questions, Lonni Sue did beautifully here. "This is the only area so far," Gregory said, "where she's at the same level as the age-matched controls"—that is, people of about the same age and with similar expertise in art.

With no clear ideas about how to move forward in testing her visual creativity, the Hopkins team decided to switch gears and look at her musical talent. Instrumental music is obviously a creative activity as well, but in some ways it's more constrained than visual art is. Unless you're doing jazz improvisation, you read the notes on the page and you manipulate the clarinet or viola or tuba so that it makes the corresponding sounds. In a way, it's purely mechanical. With most beginners, at least, it sounds purely mechanical. Good musicians, however, like Lonni Sue before the encephalitis, add emphasis and emotion to the basic notes by subtly slowing the tempo or speeding it up, hitting some notes harder than others, alternating between louder and softer volume, adding vibrato. Aline had told the scientists that her sister could still play the viola, and play with some of the emotion she'd put into her music before the illness. This seemed like a promising thing to look at.

Before they got to the creative part, they wanted to know how much factual knowledge Lonni Sue had retained about music in general and specifically about playing the viola. As the first step in testing her artistic knowledge, they'd asked her to identify works by well-known artists. Now they asked her to identify well-known tunes—"Pop Goes the Weasel," "Happy Birthday," "Pomp and Circumstance," Mendelssohn's Wedding March, Christmas carols. For most of these, it wouldn't make sense to ask who the composer was, since even a musician is unlikely to know the answer. Instead, they asked her what the name of each tune was, and what event it's associated with. Then they played a phrase, and asked Lonni Sue to continue singing or humming the melody. She pretty much bombed on the names and associations, but she could sometimes keep the tune going.

Then, just as they had with art, they got down to more practical questions. They asked her about music theory, about the parts of the instrument ("I never knew a frog was anything but an animal," Gregory said later; in fact, it's the place where you hold the bow), about the rules of playing in a quartet ("Like who has the right-of-way," Gregory said; "I know that's not music-speak, but who leads off and such"). She did as well here as she did on the tests of knowledge about how to create art. Once again, the neuroscientists were taken by surprise. Her knowledge was detailed and specific. It wasn't totally unexpected that she could still make art and that she could still play the viola. These are largely procedural skills, which are as natural and automatic for an accomplished practitioner as walking is for the rest of us. It wasn't entirely unexpected that she'd fail to identify artists and composers, which involve declarative memory.

But these tests focused on a more nebulous border area. Lonni Sue had strong declarative memories about procedural skills, and that, Gregory said, is really intriguing. "What is it about this information? Why does she remember it so much better than she remembers everything else? Is it because she was more interested in it? She had more experience with it? She used it in some hands-on sort of way? Those are questions that we don't know the answer to." Aline thinks it might have to do with the fact that music lessons tend to include extensive conversations about how the student might improve her technique or her interpretation, but that's purely speculation at this point.

By now, Landau, McCloskey, and Gregory had been working with Lonni Sue for nearly two years, driving up to Princeton once a month, on average, to do a day's worth of testing. They still hadn't tested Lonni Sue on whether she could learn a new skill, as Milner and Corkin had with Henry Molaison in the mirror-drawing experiment. Given her much wider range of abilities, they figured that something this basic wouldn't tell them anything especially interesting. Instead, they decided to see whether she could learn a new musical piece. They knew she could read sheet music, so playing an unfamiliar composition wasn't in much doubt. What they really wanted to know was whether she could get better at it through practice.

In fact, they already had some evidence that she could. After she returned to Princeton, a family friend named Rita Asch had written a piece for her titled "Waltz for Lonni Sue." Lonni Sue had played it over and over, and she could now perform the tune with energy and confidence. Aline was convinced that her playing had improved. But even though Aline was a highly trained musician herself, her conviction didn't qualify as a scientific observation. Maybe Aline thought she could hear improvement because she hoped so fervently that her sister was getting better. It might be another instance of Feynman's admonition that being invested in the outcome of an experiment makes you the easiest person to fool. In order to test in a rigorous way whether Lonni Sue was actually learning something new, the scientists devised a novel test. They would base it on music that was entirely unfamiliar to Lonni Sue—three different pieces, they decided, so they'd be able to make comparisons. But there was a problem. None of them was a musician. They certainly knew nothing about the viola, and they didn't have a clue how to create pieces that were similar to one another, but not too similar, and also not too easy to learn.

Then Emma Gregory stumbled on a solution. It came from an

entirely unexpected direction. She was at the Hopkins gym one day working out, and idly happened to read the staff bios posted on the wall. One of the personal trainers was an undergraduate student named Joel Ramirez. He was a neuroscience major and the principal violist in the Hopkins Symphony Group. This seemed too good to be true. Gregory rushed back to the lab to find McCloskey. In order to graduate, neuroscience undergrads at Hopkins can't simply take courses. They also have to participate in some sort of research project. "Mike," she said, "you should contact him to see if he's interested in doing his project with us." McCloskey did, and found that Ramirez hadn't completed his research requirement yet. He invited Ramirez to join them.

Students don't usually say no when you hand them a juicy opportunity like this, and Ramirez was no exception. "As a neuro student," he said later, "I knew about H.M., of course." He knew about the experiments showing that H.M. could learn a new skill even without an intact hippocampus. The fact that Lonni Sue was a violist made it even more enticing. "Translating notes on a page into music is a lot more complex than mirror-drawing," he confirmed. The senior members of the research group had suspected the same thing, but his personal experience and intuition felt more persuasive.

Drawing on his expertise in music theory and in viola playing, Ramirez composed three separate pieces of music, from scratch. All three were in the same key. They were also similar in terms of difficulty, length, the number of notes of different types (half-notes, quarter-notes, eighth-notes, and such), the number of times the player had to move the bow from one string to another, the number of times the player had to shift her hand up or down the fingerboard, the number of slurs, and more.

The testing took place, as it always did, in the Johnsons' dining room in Princeton. Aline fetched Lonni Sue from her residence, and Gregory and Landau drove up from Baltimore. (McCloskey couldn't make it this time.) With them was Jussi Valtonen, a doctoral candidate in neuroscience from Finland. Valtonen had done a master's degree in cognitive science with McCloskey at Johns Hopkins back in the early 2000s, but stopped short of getting his Ph.D. because he wanted to return home. He missed his family. Then Valtonen turned to fiction, writing several novels. ("Unfortunately," Gregory said, "they're all in Finnish.") And then he decided to go for his Ph.D. after all. He'd wondered if he might collaborate with his old professor on a piece of research, and this one seemed to be a great fit: Valtonen plays guitar, and he knows a lot about music.

When the scientists arrived, Lonni Sue was sitting at the dining-room table with colored pencils and piles of graph paper spread around her, concentrating intently on creating her latest puzzle. Maggi and Aline greeted the visitors like dear old friends, with wide smiles and hugs all around. This isn't how a scientific research session normally begins, but the Hopkins crew had been there so many times already, and the Johnsons are so warm and welcoming, that it would have felt odd not to do it. At Henry Molaison's memorial service, Suzanne Corkin had spoken of his contributions to humanity, but she'd also said that "for many of us, losing him was like losing a family member." H.M. had been so cooperative, likable, and funny that the scientists had developed a real affection for him. The same was true for the Johns Hopkins team and the entire Johnson family.

Lonni Sue looked up from her work, just as she had when I'd first met her. "Hello!" she said brightly. "So nice to meet you! My name is Lonni Sue. What's yours?" She had no idea that she'd seen any of them before, even though they'd met at least two dozen times. The scientists were very familiar with the routine by now; they introduced themselves all around, as they always did. Then Lonni Sue asked, as always, "Do you like to draw?" Yes, they all liked to draw, but they assured her that they weren't as good at it as she was.

Then she asked, as she nearly always did, "Do you like to sing the alphabet?" As always, she demonstrated, improvising a song on the spot. This was extraordinary in itself: not only could she pluck a word out of the air for each letter without more than a moment of hesitation, but the tune she created on the fly was melodic, albeit a bit meandering. Sometimes she'd do an alphabet game rather than a song. It worked on the same principle, but with no melody and with two players alternating words as they made their way through the alphabet. Often, there was a theme—animals, for example, or foods. Once, I saw her do an alphabet dance: she sang out each letter (no words this time) and formed her body, as best she could, into the letter's shape. Some came out better than others. (The letter Q turns out to be pretty tough.) She couldn't stop laughing at the silliness of the whole thing. One peculiar characteristic of her alphabet songs and alphabet games was that negative words were forbidden. When an old friend from high school came up with "death" during an alphabet game, Lonni Sue gently told her no, that wasn't a good word. She'd have to pick something else.

Aline has thought a lot about her sister's fascination with the alphabet. "A friend of Lonni Sue's suggested that she uses it as a scaffold," Aline said. The sequence of letters had been burned into her memory so deeply, at such an early age, that it was purely automatic, like riding a bicycle, but even more deeply ingrained. The encephalitis couldn't wipe it away. Each letter led her automatically to the next, without her having to remember the entire sequence in any conscious way. The same was evidently true for music. The path up and down the musical scale was seared into the deepest recesses of her unconscious. If she sang the note A, it was the most natural thing in the world to proceed one step up to B or one step down to G.

When Lonni Sue was finished with her alphabet song, everyone applauded. Then Gregory explained that they were going to do some tests. "Will this help other people?" Lonni Sue asked—earnestly, as she almost always did, with her brow furrowed in concern. When Maggi and Aline had decided that making a contribution to science might wring something positive out of the tragedy, they'd begun impressing this idea on Lonni Sue. Eventually, it had stuck.

Scientists had seen something similar with Henry Molaison. He'd lost his ability to form new memories in 1953, but several decades later, during a testing session, a researcher named Edith Sullivan asked him what he thought of when she said the name "Edith." Edith Bunker, he answered. Sullivan was flabbergasted. Edith Bunker was a character on the TV show *All in the Family,* which didn't begin airing until 1971. Henry also knew that Edith's husband on the show was Archie Bunker. He even knew Archie's nickname for his son-in-law: "Meathead." Evidently, these facts, repeated over and over on the show, were ultimately able to stick in Henry's brain. These surprisingly and seemingly random memories, Corkin writes, "appeared from time to time like driftwood washing up from an empty sea, and they felt like small miracles to those of us accustomed to seeing Henry fail to remember."

It turned out that the lack of a functioning hippocampus doesn't absolutely prevent the brain from acquiring new declarative knowledge. Given enough repetition, a fact can be permanently stored and later retrieved. Aline and Maggi had already seen this with Lonni Sue. The first time they'd reminded her that her father was dead, she was visibly upset. Five minutes later, she'd forgotten all about it. She was just as upset the second time they told her. It went on and on in a seemingly endless loop, but gradually, after many, many reminders, the fact of Ed's death had worked its way into her memory. If you ask her about her father, she'll now tell you, "He died." In the same way she'd also learned that her participation in the research studies would help other people. The scientists reassured her once again that yes, it would.

They set up a video camera and sound-recording equipment. When everything was ready, Lonni Sue picked up her viola, stepped to the music stand, and peered at the sheet of paper in front of her. "Oh!" she said brightly. "It's called 'Caprice'!" (All three tunes had that name, in fact—they were called "Caprice A," "Caprice B," and "Caprice C," although only the word "Caprice" appeared on the page.) "What language is that?" she asked. "I think it's Italian,"

Gregory said. (It's French, but that's irrelevant.) "I love that word," said Lonni Sue, grinning broadly. "Did you notice that it's made up of 'cap' and 'rice'?" She was doing word searches on the fly. "That's really interesting!" Gregory said, enthusiastically. Lonni Sue scraped her way through the piece. Then Gregory took the music away and put a new sheet on the stand, the second of the three Ramirez had composed. It had the same name, but while the switch had taken only seconds—well within her brief window of memory—the distraction of having the paper pulled off the music stand and replaced with another had broken the continuity of the experience. "Oh, it's called 'Caprice,'" Lonni Sue said, not realizing she'd said exactly the same thing a few minutes earlier. "Did you notice . . ." This would go on all morning. Her audience laughed each time, not because it was such a clever observation, but because Lonni Sue was so genuinely amused and delighted.

"One thing that's just so different about Lonni Sue compared with other subjects," Gregory said later, "is that she's so excited all the time. It's not like you're bothering her to do the testing." She'd rather be drawing than taking tests, and she asks repeatedly during a session when she can go back to her puzzles. But once she focuses, she's one hundred percent there. "That's just amazing," Gregory said. "You can get control subjects who are completely neurologically intact, and they aren't as focused as she is." Lonni Sue would sometimes get frustrated with the testing, whether it involved music or art or pretty much anything, but more often she'd make a joke about it. "Sometimes that would take her into a laughing fit," Gregory said, "which would get her unfocused. But it was worth it to see. She has that really pure laughter and that smile, and it's wonderful."

Gregory had Lonni Sue play through all three caprices, one after the other. It was a struggle. Ramirez had deliberately written pieces that were difficult to play and not very melodic, so that she couldn't use her musical intuition to make an unconscious prediction of what might come next. For the same reason, Gregory had set a metronome ticking uncomfortably fast. Lonni Sue had a hard time keeping up. She would keep asking, "Do I have to play it so fast?" "Just do your best," Gregory would reply, reassuringly.

After she'd played each tune once through, the scientists began feeding her random excerpts from the first piece, "Caprice A"—harder bits, which they had her play a lot, and easier bits, which she did less frequently. They gave them to her all out of order. Sometimes they had her play the whole piece all the way through. It's how a non-amnesic might have broken down a difficult piece of music to practice, but Lonni Sue never knew she was practicing, because she never realized that she had ever seen these notes in this order before. Gregory and the others put her through two separate training sessions that day. During the morning, she played all three pieces, then practiced "Caprice A" intensively, then finished by playing all three again. During the afternoon, she played all three pieces, then practiced "Caprice B" intensively. She never practiced "Caprice C." That one was the control, which the scientists would later use for comparison.

In between the sessions, Aline and Maggi served lunch, as they usually did—this time, it was sandwiches catered from a local deli, fresh fruit, and sparkling water, although more often than not, Aline would make lunch from scratch. "I view each research session as an occasion for celebration," she said. While they were getting the meal organized, Lonni Sue went back to her puzzles, spreading out once again at the dining-room table. When Aline asked her to move so she could set the table, it was the only time during that long, tiring day that Lonni Sue showed any real annoyance. "I really have to work," she told her sister, bristling, and at first she refused to move. But she finally acquiesced, and the group sat down to lunch. Lonni Sue suggested that everyone hold hands and say grace. When the girls were young, the Johnsons had always paused for a moment of reflective silence before dinner. Lonni Sue evidently had a glimmer of memory about that custom, and somehow she'd picked up the notion of a more traditional grace. "I'm not religious at all," Gregory said later. "I'm like, 'I hope I don't say anything inappropriate.'"

An outside observer would have had trouble figuring out which ones were the scientists and which the hosts, as they all talked and laughed and caught up with one another's lives. "They definitely took us in," Gregory said. "They seemed really happy to have us."

Then it was back to a caprice, which, in case anyone hadn't realized it, was made up of the words "cap" and "rice." Lonni Sue made sure to tell them.

Two weeks later, the scientists showed up once more with their recording equipment, and asked Lonni Sue to play all three caprices again. She hadn't seen any of the pieces in the interim. She didn't remember having ever played them. At one point, she complained that they were making her play too fast. "Nobody makes you play a piece this fast when you play it for the first time," she said, plaintively. "Actually," Landau said, "you've played this quite a number of times before." You might expect Lonni Sue to have argued or gotten upset, since this was directly at odds with her own experience. She simply said, "Oh, I have?" and kept on playing. "Her awareness of her memory problems is a whole other really interesting piece of this puzzle," Gregory said. "I mean, she clearly doesn't understand how severe her amnesia is. I don't know if I've ever heard her use the word 'amnesia.' But she will regularly say, 'My memory is bad.'" She doesn't seem distressed or frustrated when you point out a lapse, unless you keep pressing her to recall something, even when she's said she can't. After the second or third time, she's liable to say, "I told you I don't *remember,*" with a touch of annoyance. But it's clear that she's upset by the pestering, not by her deficiency.

What the neuroscientists wanted to know was whether Lonni Sue would find it easier to play caprices A and B, the ones she'd practiced intensively during the first session, compared with "Caprice C," which she'd played through only a couple of times. And she did. "The long and the short of it," Gregory said later, "is that she absolutely improved on the pieces she'd practiced." This wasn't Gregory's judgment, or Barbara Landau's, or Mike McCloskey's, or Joel Ramirez's, or Jussi Valtonen's. Aside from Ramirez and

Valtonen, none of them knew enough about music performance to say anything useful, and all of them, including the musicians, were too close to the experiment to be objective.

Instead, they scored Lonni Sue's playing in two ways. The first relied on "coders," skilled amateur musicians recruited by the scientists to act as independent judges. The coders listened to recordings of the practice sessions while looking at the sheet music, and kept track of how many notes she got right in terms of pitch, duration, tempo (that is, how close the note fell to where the metronome said it should fall), and more. The second set of scores came from experienced violinists and violists. They listened to recordings of Lonni Sue playing the caprices and made their own subjective judgments of how she sounded overall. The scientists prepped these judges by having them look over the sheet music, play the pieces themselves on their own instruments, listen to a "perfect" rendition of the pieces performed by a computer, and listen to a recording of Lonni Sue playing the waltz that Rita Asch had written for her. The last item was her best post-encephalitis playing, and was designed to serve as a baseline for comparison; it helped the judges know how well she was capable of playing when the tune was simple and familiar. "If they'd compared her playing to their own, even subconsciously," Gregory said, "she might have gotten really bad ratings from all of them."

The results—tabulated, analyzed, and laid out in a paper in the journal *Frontiers in Human Neuroscience* in 2014—were clear. Lonni Sue's playing did improve with practice, even though the pieces seemed new to her every time she saw them. When the Hopkins neuroscientists tested her the second time, she still played the caprices better than she had the first time, even though she hadn't practiced them in between sessions. H.M.'s unconscious improvement in the simple motor task of mirror-drawing five decades earlier had taken Brenda Milner by surprise. Now Lonni Sue had improved at a much more cognitively complex task. "Thus," they wrote in their paper, "there was considerable room for improvement in LSJ's ability to cope with the excessive cognitive load, and this—and not motor acuity—is likely where her learning occurred."

Landau, McCloskey, and Gregory had probed two of the areas where Lonni Sue's expertise and skill were much greater than those of the average person. Despite the terrible damage to her medial temporal lobes, she could still make art, and she could still talk about how you go about making art, in impressive detail. She could read music, play the viola with real feeling, and get better at playing a new piece through practice.

Now the scientists moved on to her third area of accomplishment. They put together a test that probed her knowledge about flying. "We asked her about the parts of the airplane," Gregory said, "along with questions like: 'How do you deal with this sort of wind or that sort of wind? What do you do in a stall? Who has the right-of-way in this situation? How do you get a license?' and things like that." Maggi had helped them create the art test; Ramirez and Valtonen had been the expert consultants on the music tests. For the flight-knowledge test, Gregory recruited a pilot to come up with a battery of questions about flying. "That stuff is really difficult if you don't know anything about it," she said—the same thing she'd said about art and music.

The results were consistent with what the scientists had learned about her other areas of expert knowledge. When I spoke to them about it, they were still in the process of writing up a detailed, rigorous analysis for publication in a journal (it was ultimately accepted by *Cognitive Neuropsychology*). Gregory was happy to sum it up for me in laymen's terms. "Basically," she said, "what you see is that Lonni Sue knows stuff." She doesn't know as much as she used to, and in the case of music and aviation, at least, she doesn't do as well as control subjects who are as accomplished as she used to be. "But she's way better than most people," Gregory said. "She doesn't know everything. But she knows a lot."

Lonni Sue does know a lot, and not just about the technical details of flying. Aline asked me if I'd like to interview Lonni Sue.

I spent about an hour with her, and at one point I asked her to talk about flying. What does it look like when you're up in a plane? I asked her. "As you take off, it shrinks," she said, meaning the ground. "Then you get to see bigger pictures of clouds and the moon and the sun. But you always have to be looking, because there are birds up there and other planes, and blimps. Then as you land it enlarges again."

And what does it feel like to fly? "It's exalting. It's a rhythmic, musical thing. When you play the piano," she explained, "you have your two feet on pedals and your hands here." (She knew this because she'd had piano lessons for several years as a child, and frequently played in ensembles that included pianists.) She held her hands in front of her, poised as though she were about to plunk them down on a keyboard. "That's so much like flying, because when you fly you have your feet on the rudder pedals and your hands on the stick or the wheel. Also," she said, "you tilt back and forth when you play."

Do you tilt when you fly? I asked. "Yeah. Sometimes if the wind is blowing at you it sort of uplifts your wings, but you're still going straight ahead. Then if it's coming from the right side it wants to uplift one wing so you have to steer the rudder to make sure that you're still going straight. Then if you're going with the wind coming behind you, that's really tough. But if you have to, at some point you just have to figure out what to do. You don't want it topsy-turvy unless that's part of a thing where you're doing a design in the sky."

Then I asked her what learning to fly was like. "It was exalting," she said. "I mean it was so exciting taking lessons where you go up with a teacher and they tell you what to do. They'd be sitting next to you or behind you. Usually you start off next to a pilot. 'Pilot' like on a stove," she said, unable to resist a pun. "You learn the knack and you feel it. The first day that I got to be a soloist was so exciting," she said, using a term from music instead of flying. It made perfect sense nevertheless. "Finally, the teacher got out of the cockpit and said, Okay, it's your time to try it alone."

Was that frightening? "No, it was heavenly, literally. We just went in a loop around the runway, upwind, downwind. It's nice how the runway is set up. They have a wind sock, so you know which way you have to land, because you land into the wind. But before you land you have to figure out which way you land. There are radio stations. Not if you have a private airport, but you want to make sure that you're not landing head-on with someone else. You have to choose. Sometimes they say which way they're using right now. Then if the wind is coming from the side you have to know which way to tip your wings so you don't get skidded off." There's no record that H.M. ever delivered a mini-lecture like this. That might mean he was incapable of it—but it might mean that he didn't have anything to lecture about, because his pre-amnesic life was so prosaic.

So what kind of memory is this? It seems to be episodic, since her first solo flight happened only once, by definition, and she talks about what the teacher said to her. The quote feels generic, though, and so does her account of the solo flight. That pushes things more in the direction of semantic memory. But then, her vivid and poetic description of what it looks like and feels like to fly isn't generic at all. It's general—it doesn't describe a specific episode—but it's also intensely personal. Nobody else could have experienced flying just that way, with the sensibilities of both an artist and a musician.

It's tough to nail down exactly what category these memories fall into. It's also tough to know what category of memory allowed Lonni Sue to get better at playing the caprices. The reason is that the different types of memory Brenda Milner and Suzanne Corkin found in their experiments with Henry Molaison were almost certainly oversimplified. "The work with H.M.," Gregory said, "gave us these basic distinctions"—between declarative and implicit memory, between episodic and semantic memory. "That was incredibly important."

Those distinctions may have been too crude to capture the subtleties of human memory, however. That's not really surprising, when you think about it. When we put things in categories, we do it for our own convenience, not necessarily because nature has divided things that way. Categories are a human invention. That was made evident in another area of science recently, when Pluto lost its status as a full-fledged planet. For decades, astronomers agreed that there were exactly nine planets. Then they looked through more powerful telescopes and began finding things just a tiny bit smaller than Pluto. So were these objects planets, too? Or were they something else? If they weren't planets, why should Pluto continue to be one, since it was only marginally larger? That brought up the question of what the exact dividing line should be between a very small planet and, say, a very large asteroid. Nobody had ever come up with a rigorous definition, because it didn't seem necessary. Now that it was necessary, it turned out to be extremely difficult to agree on one. In fact, astronomers still don't agree. The problem isn't with Pluto. It's with the categories humans invented to try to make sense of nature. Just because we put things into categories doesn't mean nature is obliged to cooperate.

The same is evidently true for memory. The divisions neuroscientists originally created, based on Henry Molaison and on a handful of amnesia patients who came after him, are very roughly accurate, but the boundaries are fuzzy. Neuroscientists know that. "We need to be very careful not to attribute to the brain the categories we've invented through our own introspective experience or experience in the world, and as captured in our language," Neal Cohen, who had worked with H.M. as Suzanne Corkin's graduate student, said.

Henry Molaison could remember general facts about his life before surgery, but only two of them involved specific incidents. He remembered his first airplane flight, and he remembered the first time he smoked. Neuroscientists would call these his only episodic memories. The rest, including his recalling where he went to high school or the fact that his parents would take him on vacations driving the Mohawk Trail, were semantic memories—general facts about his own life, with no fine detail.

But where, exactly, is the line that distinguishes episodic from semantic memories? Let's say you remember that you went to the senior prom in high school. Maybe you also know that you went with this person, and had that entrée for dinner. Does that make it an episodic memory? Some neuroscientists would say yes. Others would say no, that you have to remember more details, like, I was feeling like this, and we sat here, and I had this particular conversation. "It's very complicated," McCloskey said. In some areas of brain research, he said, neuroscientists have pretty good theories that let them design experiments and help them interpret the results. "Memory is not really one of those areas," he said. "Once you get past global statements about episodic versus semantic memory, there's not a lot of specific theory there."

The same difficulty applies when you try to distinguish between declarative and procedural memory. In the oversimplified version of how the brain divides things up, Henry Molaison couldn't form new declarative memories, because he had no functioning hippocampus. He could, however, learn new skills. Therefore, the hippocampus isn't involved in procedural learning. Things like riding a bike are handled by a different memory system located in a different part of the brain. That's what the textbooks say, more or less.

As early as the 1960s, however, when testing on Henry was just getting under way, some experts on learning were pretty sure that wasn't how it worked. In their 1967 book *Human Performance,* psychologist Paul Fitts, of the University of Michigan, and neuroscientist Michael Posner, of the University of Oregon, argued that many forms of procedural learning do involve declarative memory. When you learn a new skill, they argued, you do it in three stages. The first is a cognitive stage, where you pay conscious attention to how to do whatever the skill is. When my English teacher Mr. Cook taught me to tie a bow tie, he explained the steps verbally, and I followed them consciously, trying to turn the words into physical actions. When my friend Jim taught me how to drive a car with a manual transmission, he explained exactly how to let out the clutch and press down on the accelerator to get the car moving without stalling. In both cases, I parroted their instructions internally as I tried to make my body do what it was supposed to do.

In the second stage in Fitts and Posner's model, your motor system learns to associate the verbal description of what you're supposed to do with the feeling of doing it right. And in the final, autonomous stage, you just do it without thinking. You might even forget how to describe the action verbally, as I did with tying the bow tie. You're on autopilot after that.

This certainly describes how Lonni Sue must have learned to play the viola. At first she'd had to use her declarative memory. She'd look at a sheet of music and remind herself that a dot in this location means A, a dot in that location is a B; this is a quarter-note; this piece is in the key of G minor. She was trying to master a form of written communication that was alien to her. But gradually, just like a small child learning to read words on a page, she wouldn't have to think about what a particular dot in a particular location on the sheet of paper meant.

The same thing applied to playing the instrument. "When I see a B-flat on the page," she would have said to herself, "I put this finger down on this string, and I simultaneously move my bow, pushing down with some pressure—not too little but not too much." Gradually, this, too, became automatic. For a musician, this is very much like riding a bicycle. The act of playing, like the act of riding, has moved out of the realm of conscious thought. You don't think about it; you just do it.

When you look at procedural learning this way, you realize that the tests Henry Molaison went through were inadequate to understanding what procedural learning actually is, and what parts of the brain are involved. "You may be able to execute some sort of action in a muscular sense," Nicholas Turk-Browne, a Princeton University neuroscientist who has also studied Lonni Sue, said. "But another big part of it is knowing what action to take in the first place." If you put a pencil in H.M.'s hand and told him to trace the image of a star in the mirror, he could do it, and he could improve with practice. In a more complex environment, however, where you need to remember the strategy you used in the past, or when you need to choose from several possible actions, the cognitive load is a lot heavier. It's quite likely, many neuroscientists now believe, that procedural learning involves multiple memory systems working together.

The promise of Lonni Sue's case is that it might help clarify, in ways that have never been possible before, not only what the important distinctions are between these different systems, but also how they work together to create our rich experience of the world, past and present. When McCloskey and Landau and other neuroscientists reassure Lonni Sue that studying her brain could be valuable to science and that it could help other people, they really mean it.

*Chapter 11*

# PICTURES OF LONNI SUE'S BRAIN

On a cloudless, unseasonably warm morning in October 2015, Nicholas Turk-Browne is on the lookout for the Johnson sisters. Lonni Sue is scheduled for testing at the functional magnetic resonance imaging lab at Princeton University's Neuroscience Institute. The testing is supposed to begin at nine o'clock, and other scientists have the machine booked for later that morning. It's important to get Lonni Sue in and out as quickly as possible, but Lonni Sue is pretty much impossible to rush. Turk-Browne stands outside the building's southern entrance, squinting into the morning sun, scanning the parking lot for the sisters' arrival. He can hear the shouts of students practicing some kind of sport on the adjacent athletic field.

"Oh, wait, there they are," he says. "They were taking a walk. They must have gotten here early." He steps out to greet them. Lonni Sue is warm and cordial, as always. She's not entirely sure who this tall, bearded man is, but maybe he seems familiar. Lonni Sue doesn't realize it, but this is because he's been conducting tests on her for more than two years. "It's a beautiful day to fly an airplane," she says, a bit wistfully. "I used to have two airplanes, a Piper Cub and a Cessna."

"There are too many clouds in the sky for flying, I think," Turk-Browne teases. Lonni Sue thinks he's being serious, and corrects him. "No, there are hardly any!" She's very earnest. "And the clouds are almost never loud," she says, now smiling broadly. "C-L-O-U-D," she explains, to make sure everyone gets it. "That's very true," he says. "So, shall we head in?" Turk-Browne and Aline walk ahead, chatting, while Lonni Sue falls in next to me. "Do you think I can go to the ladies' room?" she asks me. She assumes, correctly, that pretty much anyone she encounters is likely to know more about what's going on than she does. I assure her that she'll be able to.

"Are they going to examine my brain?" she asks. They are, I tell her. "And they make those maps of it, right?" It might seem impossible that she could know that this is what's going on. She doesn't recognize me, although she's met me at least a dozen times now. She couldn't tell you Nick Turk-Browne's name to save her life. Yet even without a hippocampus, she's managed to absorb and hold on to this bit of information, just as she gradually did with the information that her father had died.

Yes, I told her, as we entered the Neuroscience Institute, they will make maps of your brain. It's amazing, I say. "The maps? Or my brain?" she asks, genuinely curious. Both, I tell her. I ask if she's ever had her brain mapped before. "I might have," she says. We follow Turk-Browne and Aline into the fMRI lab, a suite of rooms that includes the chamber housing the massive fMRI scanner itself, a control room next door, a small conference room, and a bathroom, to which Lonni Sue immediately repairs. "She'll be in there for at least ten minutes," says Aline, who knows from experience. Aline greets Jiye Kim, a postdoctoral student who will be in charge of today's testing. There's a sense of palpable warmth, just as there is with the researchers from Johns Hopkins, and just as there was between Suzanne Corkin's group and Henry Molaison. All of the scientists consider the Johnsons to be friends and collaborators, not just test subjects. "Good to see you again!" Aline says. "You need me to fill out consent forms, right?"

Before she does, Aline urges Lonni Sue, who has just emerged, to divest herself of the things that need to stay outside the fMRI room. The machine incorporates such powerful electromagnets that anything metal can pose a real hazard. Keys have been known to fly across the room at high speed when an fMRI cranks up. So have chairs. (To be clear, no such mishap has ever happened at Princeton.) Metal implants within a subject's body could be even more hazardous—you don't really want to think about how. Lonni Sue has no implants, fortunately, but anything she might be wearing or carrying—watches, jewelry, pens, eyeglasses—his to be left behind. Aline is in charge of making sure that happens. "See?" she says, showing Lonni Sue the bag where she's stashed her sister's necklace, a headband, and several other items. "We'll put it in this plastic bag so you'll know where it is," she says. "Let me have your tissues."

The tissues have no metal in them, obviously, but there's always a chance Lonni Sue will have a pen tangled in the wad she always has stuffed in a pocket. "We have these special tissues you can take with you," Aline says. She hands her sister a box on which she's written "Special fMRI tissues" in ballpoint. "Okay, turn out your pockets, like this." She shows Lonni Sue a cartoon drawing of a person with her pockets pulled inside out to show there's nothing left. She's written "wings" on the drawing near the protruding pockets, which she thinks might appeal to Lonni Sue's love of aviation.

Once Aline is satisfied that her sister has no metal contraband on her, Turk-Browne has Lonni Sue hold her arms out straight from her shoulders while he "wands" her with the kind of hand-held metal detector that TSA agents use at the airport. Sometimes she'll do a pirouette afterward, her arms still extended, but not today. Then they go through a full-size metal detector, also airport-style. Turk-Browne explains what it will be like in the fMRI, since Lonni Sue doesn't remember the dozen or more times she's been in it already, and then in she slides. A neuroscience graduate student named Nayeon Kim (no relation to Jiye) will sit in a (plastic) chair next to the scanner to reassure her if she gets agitated—although she never does—while Turk-Browne joins the other Kim at the console next door. Although Lonni Sue can't remember, she's been through essentially this same routine many times since the fMRI studies began in late 2011.

The research at Princeton started as the result of a chance encounter, just as the original studies by Landau and McCloskey had. Laypeople tend to think scientific research is more organized than it often is. In the movie version of Lonni Sue's story, the scene where Robert Landau tries to get his reluctant scientist wife interested in the case would probably stay in, because it has a certain cinematic drama. But after that, the actress playing Barbara Landau would probably put out a call to the top neuroscientists in the world. "This is a once-in-a-lifetime opportunity," she'd tell them, "a chance to go for the scientific gold. We need to look at this patient's amnesia from every possible angle. Drop what you're doing and come join the research team in Baltimore at once." There might even be helicopters.

The way things unfolded in the real world was a lot more random—kind of an "I know a guy who knows a guy" sort of thing (although many of the guys in this case were female). No one could have predicted that the first scientist to start looking at Lonni Sue would hear about the case from her husband the woolen-goods merchant. No one could have guessed that the studies of Lonni Sue's viola playing would come about because Emma Gregory noticed a personal trainer's bio taped to the wall of a gym. The haphazard nature of this kind of research has a long pedigree. Back in the 1950s,

William Scoville just happened to read a paper by Wilder Penfield describing how some brain-surgery patients developed amnesia. He wrote to Penfield, who sent Brenda Milner down to test Henry, which turned Henry into H.M., the most famous amnesia victim in history. If Scoville hadn't read the paper, who knows what would have happened. Sometimes scientists have a casual conversation and realize that their interests align. They decide to collaborate. If they hadn't had the conversation, an entire line of research might never have come about, or at least would have been delayed.

That's how Turk-Browne came to work with Lonni Sue. Shortly after Landau, McCloskey, and Gregory began their research project, he came down to Baltimore to give a talk on his own work. Turk-Browne is especially interested in the brain's visual system and in statistical learning. The latter is a process by which we unconsciously pick up an understanding of the world by detecting patterns and regularities in our environment. That in turn helps us make predictions about what to expect, and alerts us when something out of the ordinary turns up, in case we need to deal with it. "Just by exposure to something," he told me on a visit to his office in Princeton, "you're extracting information, without doing so intentionally."

Here's a relatively trivial example from my own life. A couple of years ago, my stepson and his family moved out to central Pennsylvania. I've made the three-hour drive there and back many times now. At first, I was totally unfamiliar with the route. I simply went where the GPS told me to go. I had no idea of what I'd see when I rounded the next curve, or the one after that. Every mile of the trip was a novel experience.

Gradually, however, I began to anticipate more and more often what was coming next. When I see the U-Rent-It place on the right, I know that the restaurant built in an old power station will be coming up shortly on the left. When we start on that long downhill that curves to the right with the wide valley on the left, we'll soon come to that shopping mall with the Home Depot. I never set out to memorize these things, but my brain picked up on them anyway.

It's at least moderately useful because it lets me answer the age-old question "Are we there yet?" When I see a wide stream running along the right side of the interstate for a couple of miles, I know we almost are.

That's the kind of thing we do all the time as we navigate the world, in every meaning of the word. "Many people think this is also part of the way young infants learn language," Turk-Browne said. The idea is that some combinations of syllables occur more often than others, and that babies' brains take advantage of this statistical frequency to learn words. The infant gradually learns that she hears the sound "mama" more often when she sees a specific face and hears a specific voice than when the face and voice are absent.

Turk-Browne and a graduate student named Anna Schapiro had been using Princeton's fMRI to try to understand which parts of the brain are involved in statistical learning. A conventional MRI uses powerful magnets to flip the protons in your body's hydrogen atoms into alignment. When the magnets turn off, the protons flip back, sending out a tiny ping of energy. The MRI detects those pings. Since hydrogen atoms are part of every water molecule and fat molecule in your body, the pings can be assembled into images of the soft tissues that X-rays can't see. An MRI can map out the structure of your brain in exquisite detail.

An fMRI isn't quite the same as an MRI. It uses the same machine but a different technique, one that's sensitive to the oxygen-carrying hemoglobin in your blood. It can measure blood flow as it changes over time. The firing of neurons in your brain takes a lot of energy. When blood flow increases in a particular region, it tells neuroscientists that neurons are firing there. An MRI is good for showing what your brain looks like, from its outer surface to its core. An fMRI is good for showing which parts of your brain are active while it's working on different sorts of mental tasks. An fMRI has its limitations: blood flow can show which parts are active, but it doesn't tell neuroscientists precisely what those parts are doing. Contrary to some breathless popular accounts, you can't use fMRI to see precisely what someone is thinking, or catch a person telling lies, in the sort of open-ended way that would stand up in court. But it can tell you which parts of the brain are working hardest at a particular time.

Before Lonni Sue came along, Turk-Browne had been investigating statistical learning with non-impaired subjects. "What we do," he said, "is to show people a series of objects—shapes, or scenes, or something like that, that they've never seen before." He asks the subjects to make judgments about each image. For example, there might be images of places. When an image flashes on a screen, the subject has to say whether it's an indoor place or an outdoor place. It's similar to, but not exactly the same as, what Lonni Sue was doing on that October morning. "They're making this judgment about every picture," he said. "They think this is what the experiment is about. It's not."

In addition to the images of places, Turk-Browne scattered images of faces. In a hypothetical example, he might use his own face, and one of the places might be an indoor place—say, his office. The order of images seems to be random, but it's not. "After you see my face in the sequence you would see a picture of my office, not every time, but more often than you'd expect by pure chance," he said. And, in fact, the subjects learn to associate the face with the location. You can tell this without ever asking them about it: you can see them identifying the office as an indoor place faster after they see his face than they do after seeing someone else's face. "What's amazing," he said, "is that because people are performing this other task of distinguishing between indoor and outdoor, and because we don't say anything about these statistical relationships, they have no awareness that they've learned anything." They're often surprised when you tell them afterward. They can't believe it.

The subjects in Turk-Browne's tests learn something without having any conscious, explicit memory that they've learned it. According to the simplest models of how memory works, the hippocampus, which is involved with conscious memory, shouldn't be active. Statistical learning is an unconscious process that conventional theory suggests should take place entirely within the cortex. But Turk-Browne and his colleagues can see in the fMRI images that the subjects' hippocampi are active during statistical learning. This doesn't mean that the hippocampi are necessary for statistical learning, however. Maybe hippocampal activity is a by-product, somehow triggered by the process of statistical learning but not an integral part of it. "One of the ways you can test this," he said, "is to look at people who don't have that part of the brain. If they're unable to do something, or they're worse at it, or if they do it differently from people who do have an intact hippocampus, that's evidence that activity in this part of the brain is fundamental. It is not just something that happens as a by-product."

Unfortunately, he didn't have someone without a hippocampus to work with. In fact, he said, he hadn't even thought about looking at a patient with bilateral hippocampal damage, except hypothetically, simply because they're so rare. "Maybe you could find some in larger cities," he said, "but I never imagined there would be someone like this in the Princeton area." When he gave his talk at the cognitive science department at Johns Hopkins, he learned that his imagination had been insufficiently bold. After he was finished with the lecture, Barbara Landau and Mike McCloskey came up to him and told him about Lonni Sue. Wouldn't it be great to look at her for his statistical-learning experiments?

This was, so to speak, a no-brainer. Of course, she'd be ideal. He could also see immediately, as Landau and McCloskey had seen, that working with Lonni Sue would be a chance to look at all sorts of questions about learning and memory outside of his own project. Princeton, unlike the other seven Ivy League universities, has no medical school, and thus no associated hospital. (That is, if you don't count the hospital featured in the TV series *House.* The exterior shots of the show's fictional Princeton-Plainsboro Hospital are actually images of the Princeton student center.) In 2000, however, Princeton became the first nonmedical facility to buy an fMRI, purely for neuroscience research.

The fMRI lab was originally located in Green Hall, the old psychology building on campus. Then it was relocated to a small, nondescript one-story building, near the university's football stadium, that was once used for physics experiments. The physics department vacated the premises decades ago, although the building is still identified as Elementary Particles Lab West on the campus map. Then, when the university's new Neuroscience Institute building was finished, the fMRI was moved there. Starting in mid-2011, Aline and Maggi began delivering Lonni Sue to the scanner once every couple of months so Turk-Browne could slide her into the machine and image her brain. During the first session, he brought his colleague Sabine Kastner along—in another coincidence, she is the woman whose course Aline had loved so much when she was in her neuroscience-immersion phase. Kastner also studies the brain, but where Turk-Browne was trained as a cognitive neuroscientist, Kastner is an M.D. who went on to get a Ph.D. in neurophysiology—the brain's anatomy.

"I went there for the scan," she said, "to see how it would go. We're not a clinical facility." That is to say, the lab looks almost exclusively at brain function in people without any neurological pathology—volunteers, most of them students. The scientists here don't normally scan patients with brain disorders, so they don't have to deal with difficult subjects. H.M. had been almost invariably cheerful and accommodating, but he went through a period, Suzanne Corkin said, when he was very angry. He assaulted some of his caregivers at one point. Clive Wearing had been almost frighteningly agitated. "The situation with Lonni Sue," Kastner said, "was unpredictable." It was crucial to have someone present who was used to dealing with actual patients. Kastner was right to worry, but it turned out to be unnecessary. From the beginning, Lonni Sue was perfectly happy to take part in the scans. Sometimes she'd hum to herself while the experiments were going on. Sometimes she'd do a little dance, lying flat on her back. The scientists had to tell her to stop, since the slightest motion blurs the fMRI images. Sometimes she'd fall asleep.

On Lonni Sue's very first visit to the Princeton scanner, in 2011, Turk-Browne and Kastner took her into the control room after the session was over to show her what it looked like. She asked, "What are you doing?" Turk-Browne told her, "Well, we're taking pictures of your brain."

"Can I see a picture of my brain?"

"Sure!" he said. He took her over to the computer and pulled up an MRI image. She pointed to a black splotch in the center of the whitish image of the brain. It was the dead zone that the virus had created in the place where her medial temporal lobes had been.

"What's that?" she asked. Turk-Browne responded, diplomatically, "Well, that's a part of your brain that might be full of fluid."

"Oh, that's very interesting," Lonni Sue observed. She looked away for a moment, distracted, then looked back. She again pointed to the black hole in the image and said, in exactly the same tone she'd used a few seconds earlier, "What's that?"

*Chapter 12*

# FALSE MEMORY

I can still see, and hear, and feel the assassination of John F. Kennedy more than fifty years later, exactly as it unfolded, as though it were yesterday. It's what psychologists call a *flashbulb* memory, something not just vividly memorable but also shocking—a memory with a strong emotional component. I was in fifth grade in the fall of 1963, but on November 22 I was sick, so I stayed home from school. I was lying in my parents' bed, where my mother could keep an eye on me. The radio on top of my father's bureau was tuned to WOR-AM, a station in New York City with a mostly talk format before talk radio had anything to do with right-wing political rants.

The program my mother was listening to featured a homey

husband-and-wife team, broadcasting from their New York apartment, talking about recipes or Broadway shows or something. Abruptly, an announcer broke in. "Ladies and gentlemen," he said, "we have reports out of Dallas, Texas, that shots have been fired at the presidential motorcade. Stay tuned for more details"—something like that, anyway. Then it was back to the show for a few minutes. And then the announcer broke in again, his voice much graver. "Ladies and gentlemen . . . the president is dead." Immediately, the audio switched to somber classical music. My mother started to cry. I was profoundly embarrassed. "I'll go get ready to go to the doctor," I said, looking for an excuse to get out of the room. "We're not going to the damned doctor!" she said, with a vehemence that embarrassed me even more. In the end, we went anyway.

It's as vivid in my mind now as it was on November 22, 1963. Except that it couldn't have happened that way. A few years ago I went back to look up the WOR transcript from that day, and I had all sorts of details wrong, including the sequence of announcements and the funereal music. I never would have believed that my memory could be at once so rich and so inaccurate.

Elizabeth Loftus has no trouble believing it. Loftus is an experimental psychologist at the University of California, Irvine. She's made a career of debunking the idea that memory is necessarily a reliable guide to what happened in the past. Back in the 1970s, she pioneered the idea that memory is not like a tape recorder. It doesn't faithfully transcribe what happened for later playback. Even the most vivid of autobiographical memories are constructed of bits and pieces—sights, sounds, sensations, emotions—many or most of which might be accurate on their own, but which can be contaminated and distorted by what's in our minds when we recall them. "It's not just forgetting," she said during a phone conversation. "There's obviously that, but there's acquiring misinformation from other sources and places, and adopting it as your own memory." You might also, she said, come to your own conclusions about what happened, or what might have happened, and sometimes those, too, are incorporated into your own memories.

When you're reminiscing about where you were when Kennedy was shot or, for my parents' generation, when Pearl Harbor was attacked, it probably won't cause much harm, aside from boring young people for whom both Kennedy and Pearl Harbor (and even 9/11 by now) are events from prehistory. When you're testifying as a witness in court, it can be a lot more problematic—when you're absolutely sure, for example, that the man in the defendant's chair committed some awful crime. Eyewitness testimony is enormously powerful, but Loftus has helped show that it's also frequently wrong.

She came to do the research that demonstrated this, she said, when she was having lunch with a cousin, during graduate school at Stanford. At the time, Loftus was studying semantic memory, and how it's stored and categorized in the brain. In one semantic-memory test, for example, she asked one group of subjects, "What is the name of a fruit that's yellow?," and told another group, "Give me the name of a yellow fruit." On average, the first group came up with the answer—which was nearly always "banana," naturally—a quarter of a second faster than the second group. The result suggested that people sort this kind of information by category rather than by attribute, she said. If the word "yellow" comes first, it doesn't give you enough information to pounce on the answer when you hear the word "fruit." If "fruit" comes first, you're better primed to take the next step.

Loftus's cousin, an attorney, found it absurd that she was wasting her time (and the government's money) on something so useless. Loftus had to agree. She realized she would be much happier trying to do something with a more practical application. Loftus had a side interest, although no training, in the law. It occurred to her, she said, that research into the memories of eyewitnesses would be a natural direction for her to go in. Her first experiments along these lines involved witnesses to traffic accidents, mainly because the U.S. Department of Transportation had funding available for accident-related research. She and a colleague, John Palmer, showed people films of car crashes, then asked questions about what the subjects had seen. With some subjects, they asked, "About how fast were the cars going when they smashed into each other?" With others, they used the words "collided," "bumped," or "contacted" instead of "smashed."

On average, the subjects who heard "smashed" came up with higher speed estimates than those who got more mildly worded questions. Then they asked subjects whether they remembered seeing broken glass. The "smashed" group was more likely to say they saw glass than the others. In fact, there was no broken glass. Simply by changing the phrasing of a question, Loftus and Palmer managed to alter people's memories.

Over the next few years, Loftus did a long list of follow-up experiments that showed how easily false memories can be implanted in test subjects. For example, she convinced people that they'd had the frightening experience of being lost in a shopping mall when they were young. She'd get parents or older siblings to feed the subject some bare-bones facts she'd fabricated about the event, which made it seem plausible. The mall really was someplace the family would have gone, a purely fictional elderly man who "found" the crying child was a believable character, and so on. Not only would many of the subjects start to "remember" the incident; they would often add details that family members hadn't fed them—the name of the specific store where it happened, for example.

In another series of experiments, she and several colleagues convinced subjects that as young children they'd shaken Bugs Bunny's hand during visits to either Disneyland or Disney World. Not only that: the scientists also got the subjects to believe that Bugs or another character had done something slightly weird and icky—licking them on the ear, for example. Again, the subjects took on these memories, calling up details of the incident that the scientists hadn't given them. In this case, the memories weren't just false: they were impossible. People in cartoon-character costumes do roam the Disney parks, greeting children and shaking their hands. But Bugs Bunny is a Warner Bros. creation, not a Disney character. He couldn't possibly have been at a Disney theme park.

Loftus, along with other psychologists who were inspired by her work, went on to implant more and more preposterous memories. "She convinced people," wrote *Slate*'s William Saletan in an award-winning 2010 profile, "that they had nearly choked, had caught their parents having sex, or had seen a wounded animal after a bombing"—or, even more implausibly, had seen people who were possessed by demons.

That last item will have a particular resonance for those who remember the satanic ritual abuse panic of the 1980s, and the child sexual abuse panic that went along with it—most notably, in the latter case, at the McMartin Preschool in California. The accusations against McMartin teachers and administrators were, to put it plainly, nuts. The children had not just been abused en masse, according to prosecutors who brought the case to trial, but had also been forced to go into underground tunnels and have sex with animals, along with other wildly improbable things. After a seven-year trial, the charges were dismissed. It turned out that the children in this case had almost certainly had the memories of abuse implanted by psychologists. The therapists had become convinced, based on next to no evidence, that something awful had happened, and "helped" the children remember. It wasn't necessarily that the psychologists were deliberately trying to create false memories, only that their interview techniques inadvertently made it happen.

The same was evidently true for several widely publicized cases of satanic rituals, and also some accusations of murder or sexual assault. Prosecutors alleged that the witnesses' memories of these awful events had been "repressed" for decades, then "recovered," under the guidance of presumably well-meaning therapists. Loftus has testified as an expert witness in scores of criminal trials—not to claim that a witness necessarily is recalling a false memory, but that it's remarkably easy for false memories to be implanted in general. Juries haven't always bought it.

It's surprising to learn that one's own vivid memory of an event, like my memory of the JFK assassination, could be contaminated by information, real or invented, that entered the brain years later and got mixed in with the event itself. It's even more surprising that such a memory could be created out of whole cloth in such a way that it feels absolutely real. Still, recollections by very young children, or by adults of things that happened in early childhood, are usually pretty murky and impressionistic to begin with. It doesn't seem completely crazy that they could be altered yet still appear real. It's much harder to believe that you could create false memories of things that happened later in life.

In a series of more recent experiments, however, Julia Shaw, a Canadian psychologist on the faculty of the University of Bedfordshire, in the U.K., has shown that it's entirely possible to do just that. The results were surprising for three reasons. First, memories from adolescence tend to be a lot richer and easier to recall than memories from early childhood, so it would seem less likely that you could sneak a false one in. Second, the subjects in this experiment were college students; the false memories would be from just a handful of years in the past. "My colleagues said, 'There's no way you're going to convince young adults that something happened to them five years ago,'" Shaw told me. "They were sure it could only work if the event happened when they were six and now they're twenty-six or something like that."

The third surprise was that the memories Shaw and her colleagues implanted weren't stories with happy endings, like the lost-in-a-mall experiment, or stories—even slightly weird stories—of meeting cartoon characters at Disneyland. Like Loftus, Shaw was interested in false memories in a legal context. So along with her colleague Stephen Porter, she managed to convince college students that they'd committed a crime—theft, assault, or assault with a weapon—during early adolescence. In fact, the students had never had any run-ins with the police, let alone committed any crimes. Even Shaw herself didn't expect this to work.

Shaw's study involved sending questionnaires to the students' primary caregivers (with the students' permission), asking them, among other things, whether the subjects had experienced any of six emotionally upsetting events during adolescence—the three crimes described above, along with being in an accident, being attacked by an animal, or losing a large amount of money. If they'd had any of those experiences, or even any police contact, the students were ineligible to go on with the testing. The caregivers did have to come up with at least one highly emotional event from the students' childhood, however.

In the second phase of the experiment, researchers sat down with the students and talked with them about what the caregivers had reported—the real emotional experiences, along with one of the false ones, presented as real—and urged the subjects to remember both. Remembering the false experience was more difficult, naturally, but the scientists encouraged the students to keep trying. They also offered hints, in the form of "facts" from the made-up crimes—but only sparingly. When the students asked for more details, the scientists would say, "I can't give you more, they have to come from you." If the subjects still couldn't remember, the researchers said, "That's okay. Many people can't recall certain events at first because they haven't thought about them for such a long time." The idea was that if these experts were telling the students that the memories of police run-ins were real, and could ultimately be retrieved, the students would presume it must be so. To add to the scientists' credibility, the bookshelves in the examination rooms were stocked with books on memory, to subtly remind the students that these were legitimate memory researchers. Ultimately, about seventy percent of the students came to remember, utterly falsely, that they had committed crimes.

The implications are pretty obvious: given the right sort of questioning by an authority figure, an innocent person might confess to a crime, and even believe that he or she actually committed it. "If we could generate these richly detailed false memories of criminal events," Shaw said, "then they could also be generated accidentally in therapeutic or police environments." The kind of research documented in the study, wrote Shaw and Porter in their paper describing the experiment, "is essential in the quest to help prevent memory-related miscarriages of justice."

In the case of flashbulb memories, nobody is deliberately implanting false information into our brains, obviously. Yet these eyewitness memories of major events are often unreliable as well. What's evidently happening is that our actual memories of what we saw, heard, and felt are diluted by conversations we had with others afterward, and by the coverage we saw on TV. A series of studies on memories of the 1986 *Challenger* Space Shuttle explosion, for example, showed that what people remembered in interviews shortly after the disaster was different from what they remembered, equally vividly, months later. It's the same with 9/11. Maybe you saw the towers fall in real time on a live television broadcast. Or maybe you watched a replay a few hours later, and only think you saw it live (but I did see it, I swear). Those kinds of inaccuracies creep into ordinary autobiographical memories as well; the difference is that your confidence in flashbulb memories tends to be stronger.

The term neuroscientists use for this alteration of autobiographical memories as we pull them back into consciousness and then store them again is *reconsolidation*. Everyday experience is consolidated into long-term memory—that's at least in part what the hippocampus does. But when you retrieve a memory, it can be updated with new information, then reconsolidated back into memory. This isn't a flaw in our memory systems; it's a necessary part of making sense of the world, of understanding what's related to what. The Princeton neuroscientist Ken Norman likes to use the example of penguins. "Imagine that for your whole life you've only seen birds that fly," he told me as we sat in Small World Coffee, a local café where you can spot students, professors, and even the president of Princeton University. (The late mathematical economist John Nash, whose life was fictionalized in the book and movie *A Beautiful Mind,* used to hang out here as well.) "And then you see a penguin."

Since penguins don't fly, he said, your brain could create a whole new network of neurons to represent this exception to the rule. Or it could update the network that represents birds with the tag "don't fly"—but that would overwrite the important fact that most birds do fly. "What you really want to do is . . . grow a sort of a new penguin representation that has some overlap with birds, but not too much." You don't want to park the new representation right on top of the other one, in other words, because you don't want to erase the old one. But you also don't want to park it somewhere else entirely, because then you'd fail to grasp the subtler point about the world that penguins can be birds and non-fliers at the same time. "You want to find the right level of overlap," he said, "and the brain is very good at doing that."

But there's a trade-off. Our brains' skill at updating memories with new information means that the stories I tell about JFK or about having seen Princeton grads from the 1800s marching in the university's annual alumni parade (mathematically possible, at least, since I first went to the parade as a child, in 1961) are almost certainly wrong in subtle ways. It means that eyewitness testimony to crimes isn't necessarily as solidly reliable as we assume, especially since trials happen months, or even years, after a crime is committed. If a witness talks about his or her memory of a crime with investigators or with friends during that time, the original memory can be updated with facts or feelings that weren't part of the initial experience. As our brains evolved over millions of years, it was less important to our ancestors' survival to remember specific incidents accurately—they didn't have to testify in court—than it was to develop a general understanding of how the world works.

The brain's tendency to create false memory can cause real harm to other people when witnesses misremember what they saw. But it can also be therapeutic when memories are so painful that they're debilitating—in post-traumatic stress disorder, for example, where terrifying (real) memories of abuse or combat rise to the surface, forcing victims to experience the trauma over and over. In a series of experiments beginning in the late 2000s, Daniela Schiller, a neuroscientist at New York University, showed that people can learn to tone down the emotional charge of a traumatic memory. In one very simple experiment, for example, she and her colleagues taught subjects over the course of an hour or two to feel anxiety when they saw an image of a colored square, by zapping them with a mild electric shock every time the square appeared.

Then, a day later, they showed them the colored square again. The anxiety returned, but this time, there was no shock. The scientists displayed the colored square over and over, without the shock, and the subjects' anxiety gradually subsided. The day after that, they brought the subjects back in and showed them the square yet again. No anxiety. What had happened was that the original, anxiety-producing memory was retrieved, modified with new information (that is, that the square was harmless), then reconsolidated into storage.

This isn't quite as dramatic as what happens in the movie *Eternal Sunshine of the Spotless Mind,* in which a psychologist erases Jim Carrey's and Kate Winslet's painful memories of each other after their relationship ends. (Ken Norman assigns his students to watch this film.) But along with other, more sophisticated experiments, it shows how the malleability of memory can potentially be a very good thing. Loftus has done her own research along these general lines, investigating whether it might be possible to change people's behavior—getting them to prefer healthy to unhealthy foods, for example—by manipulating their memories of what they like and what they don't. She and her colleague Giuliana Mazzoni have shown that this concept, which they call *expert personalized suggestion,* or EPS, can make subjects "remember," falsely, that they once got sick from eating pickles or hard-boiled eggs, and develop an aversion to these foods. There's a problem, though: for the technique to work permanently, the subject has to be convinced that the bad memory is real, which means a therapist has to lie (in the experiment, the test subjects were told the truth after the testing was over). The neuroscience behind EPS is sound, although the ethics are obviously questionable.

Lonni Sue doesn't remember the JFK assassination, or the murder of John Lennon, which happened not far from her New York apartment, or the *Challenger* explosion, or any of a dozen events that people of her generation remember as flashbulb events. Somewhere deep in the recesses of her damaged memory, however, she evidently does have a flickering recollection of 9/11. Maybe it's because she was a pilot. It comes out as a barely recalled fragment, pulled to consciousness with the help of Mike McCloskey in a conversation recorded on video. McCloskey begins by asking, "Do you know what 9/11 refers to, or September 11?" No, she says, it doesn't ring a bell. McCloskey persists, gently encouraging her to keep trying, but carefully avoiding giving her any clues. "Think about it," he says. "Does anything come to mind?" No, nothing. He asks her to try again.

After a moment, she asks, "Did something bad happen?"

"Keep going, what else?" says McCloskey. Lonni Sue asks if it might have been an earthquake. No, keep going. Was it a war? McCloskey's words and tone remain neutral. He's giving nothing away. Then, after a pause, Lonni Sue asks, in a tone that suggests that she's still simply guessing, although what she says makes it clear that she isn't:

"Did a plane crash into a building?"

*Chapter 13*

# CHALLENGING THE CONVENTIONAL WISDOM

During a first set of experiments at Princeton, the scientists had tested Lonni Sue's capacity for statistical learning—for detecting patterns in the world around her. They did the test with scenes, for example, to see whether she'd unconsciously learn that certain scenes were more likely to show up right after others flashed onto the screen. They also tested her on shapes, syllables, and tones, presenting them in what seemed to be random order. In fact, there was a subtle underlying pattern to them, which normal subjects would pick up unconsciously by means of statistical learning. Lonni Sue couldn't do it. "She's unable to learn these patterns in the way normal people do. In fact," Turk-Browne said, "she performs no better than random chance in about eight experiments that we've done with her." The tests on normal subjects showed their hippocampi lighting up with blood flow while they were doing statistical-learning tasks. Lonni Sue had no hippocampus to light up, and couldn't learn in this way. Turk-Browne had wanted to know whether the hippocampus was incidental to statistical learning or essential. This experiment showed that it was essential.

Once they'd finished those tests, Turk-Browne and his colleague Sabine Kastner wondered, just as the Johns Hopkins group had, what else they could learn by working with Lonni Sue. The experiments Turk-Browne and Jiye Kim were doing on that warm October morning (Kastner couldn't make it this time) were based on one of the ideas they'd come up with. They were looking at adaptive learning, a technique the brain uses to pay extra attention to new things that appear in the environment and less attention to what's familiar. We're constantly barraged with sights, sounds, and other sensations, and if we paid equal attention to all of them, we'd be constantly overwhelmed. Our evolutionary ancestors wouldn't have survived long if they couldn't tune out everyday sights and sounds so they could be alert to the unfamiliar and potentially dangerous. Adaptive learning is how the brain teaches itself to do that. It's what lets you ignore the cars whizzing by as you stroll down a sidewalk, for example, but leap out of the way when one of those cars suddenly veers toward you.

It's an open question in neuroscience whether the hippocampus is necessary for adaptive learning. This made Lonni Sue the ideal subject to test the proposition, just as she had been for testing statistical learning. The neuroscientists decided to look at adaptive learning in the visual system, an area where both Turk-Browne and Kastner were especially knowledgeable. Most of us assume that our vision works like a video camera: the eye takes in what we're looking at, converts it into some kind of electrical signal, and beams it to the brain, where we view it in real time and maybe store it for later. In fact, it's vastly more complicated than this—so complicated that neuroscientists are still working to understand the process.

Without some sort of visual processing, however, the world would be impossibly confusing. On its own, the eye can't discern shapes or colors. A particular object—a coffee cup, say—looks entirely different from one angle than it does from another; as far as the eye is concerned, they're two different things. When you look at a face, your eye can't tell by itself where the face ends and the trees or the bookcase or whatever in the background begins. That's true for any object: the eye doesn't know where the boundaries between things lie, or whether one object is in front of or in back of another. And our retinas only detect three specific wavelengths of light—blue, red, and green. The rich variety of colors we actually experience are created inside our brains from those simple inputs.

In order for the scene your eyes take in to be comprehensible, neurons send information caroming around the brain to at least thirty visual-processing areas, interconnected by hundreds of neural pathways, which beam information back and forth to turn the chaos in the visual field into some sort of order. The very simplest level of processing happens in a cluster of neurons known as V1, located in the occipital lobe, at the back of the brain. V1 takes raw input from the retina and looks for patterns in the light that enters our eyes. Where are the edges of things located? Are the edges straight or curved, and, if they're curved, to what degree? What are the relative strengths of red, green, and blue light at each point in our field of view?

Two streams of signals emerge from V1, bearing all of this information and more—the dorsal stream, which travels forward across the top of the brain, and the ventral stream, which moves along the bottom. What happens at each of the multiple processing areas located along the two streams is too horrendously complex to describe in less than a full-blown treatise. Even then, neuroscientists don't fully understand how the whole thing works in detail. To oversimplify for me dramatically, Jiye Kim explained that the ventral stream, which is also known as the "what stream," is more interested in shapes, and in keeping track of how the same object can look different when it's seen from different angles. The dorsal stream—the "where stream"—is more interested in where those shapes are located in space.

A patient known as D.F., studied by Melvyn Goodale, of the University of Western Ontario, in 1991, demonstrated this difference very dramatically. As a result of a brain injury in a part of her ventral stream known as the *lateral occipital complex,* also known as the *object-selective cortex,* she couldn't recognize or identify objects. "There's no shape information in her visual world," Kim said. "When you tell her there's an object on the table in front of her, like a mug or a notebook, and ask what it is, she can't tell you." But if you ask her to guess what color the object is, she gets it right. If you ask her what it's made of—metal, plastic, wood—she also gets it right. If you ask her to grab it, she can do that, too. "She doesn't go like this," said Kim, reaching out aimlessly toward my digital recorder. "She goes like this." Kim grasped the recorder without hesitation.

D.F.'s condition is reminiscent of Dr. P., the title character in Oliver Sacks's bestselling book *The Man Who Mistook His Wife for a Hat.* As Sacks describes him, Dr. P. could identify neither objects nor faces, which is why he attempted to take his wife's head off her shoulders, thinking it was a hat. Dr. P. also had trouble identifying shapes, but only sometimes. Sacks never gives a diagnosis of where in Dr. P.'s brain the problem lies. "Sacks isn't a scientist," Barbara Landau once pointed out to me when I brought up his name, and it's true. He is a neurologist, a medical doctor who treats illnesses of the brain and nervous system (or he was; he died in August 2015, of a brain tumor). D.F. and Dr. P. both suffered from a form of visual agnosia, *agnosia* being a general term for an inability to process sensory information; it can apply to sights, sounds, smells, and even the sensation of pain. Since agnosia can't be treated, it's not especially relevant to a doctor what the cause is. Sacks is brilliant at describing the victims of neurological illness and injury in a beautiful and sensitive way. He seems much more interested in the patient's experience than in the underlying science—which is precisely why his writing is so compelling. Aline and Maggi actually approached Sacks after a talk in New York, and persuaded him to come to Princeton to meet Lonni Sue. He spent several hours with her, but in the end he didn't write about her—perhaps because he'd already written about Clive Wearing.

D.F. and Dr. P. couldn't recognize objects because they had suffered damage along the ventral stream, but since their dorsal streams were intact, they could still identify where objects were in space. To illustrate what the dorsal stream does, Kim laid a green, cylindrical, cloth-covered case decorated with cartoon kittens, about eight inches long and an inch in diameter, on her computer keyboard. The image of the case entered my eyes, where it was converted to an electrochemical signal that traveled almost instantaneously to the visual center at the rear of my brain, then began its journey through many levels of processing so that I could ultimately make sense of it. Then she moved the case from the keyboard to the desk, about a foot away, and placed it at a slightly different angle. The image entered my eye again, but as far as the dorsal stream was concerned—the "where stream," that is, which is concerned with location—it was an entirely different object the second time.

With only a dorsal stream, I could have reached out and grabbed the case both times, but I would have had no idea what I was grabbing. Even with my ventral stream fully operational, I still wouldn't have known, at least not for the first millisecond. The raw image impinging on my retina was pure gibberish—an incoherent mix of light and color and motion. In order to make sense of it, the visual system has to put it through many levels of processing, figuring out where one object ends and another begins, what's in the foreground and what's in the background, and so on. "If V1 were all we had to go on," Kim said, referring to the very first processing step after light hits the retina, "we'd be confused all the time. V2 gets it better." In V2, the second step, she explained, "you have neurons that are sensitive to whether two edges belong together or not. Something like this." She drew two overlapping squares on a whiteboard, the one in front partially hiding the one behind. Your eye sees a single, continuous boundary, sort of like a bow tie in shape, but V2 is capable of knowing that it's two objects—that the boundary isn't really continuous.

The processing goes on up the chain of complexity, not in a merely linear fashion, but with different processing centers feeding information back up and down the line, in a dizzying array of multichannel neuronal conversations. Nevertheless, after just a few thousandths of a second, a fully meaningful image comes out at the other end.

That's only the beginning. Once that image has formed, the brain has to decide what to do with it. Is the object something novel, or is it familiar? That's where adaptive learning comes in. If it's novel, the brain wants to sit up and take notice. Maybe the object is important somehow. Maybe it's dangerous. If you look at the brains of people with an intact object-selective cortex, you can see that its neurons fire vigorously when you show them something new. If you show them the same thing a second or third or fourth time, the neurons fire less strongly each time, because the brain doesn't need to expend a lot of energy to figure out what it is. This phenomenon, Kastner said, is called adaptation. The conventional wisdom in neuroscience, she said, is that the medial temporal lobe, where the hippocampus lives, supplies information about what's familiar and what isn't.

Once an object or a scene is familiar, you stop paying attention to it, even though it's lodged in your memory. "Driving a familiar route is a great example," Kim said. Say you're driving but you're talking to a friend in the passenger seat the whole time, she said. You get to where you're going without ever being aware that you turn left here, turn right there, and so on. "You don't have to think about it," she said. "You just know."

Still, you can probably picture the route if you need to, and if something significant does change—if a prominent new sign goes up or a building is pulled down—your brain will register the novelty. I used to walk to work up Seventh Avenue in New York City every morning for fifteen years. None of these walks was particularly memorable. In fact, I can't recall a single, specific event that happened on any of them, except for one: the time I saw someone slip on a banana peel. I thought it happened only in cartoons. Seeing it in real life was so surprising to me that I can still close my eyes and picture it, twenty years later.

With most of her medial temporal lobes gone, Lonni Sue, according to the conventional wisdom, shouldn't have this adaptive response seeing novel objects or scenes. The experiment Nick Turk-Browne and Jiye Kim are doing with Lonni Sue on this gorgeous October morning is an attempt to determine whether that's true. In this test, just as with the earlier experiment on statistical learning, they'll show her images of various indoor and outdoor scenes. Lonni Sue's job is to press a button when she sees an indoor scene but do nothing for an outdoor scene. As in many of Turk-Browne's experiments, the button-pressing has nothing to do with what they're studying. It's just a trick to keep her paying attention. What's really going on is that the scenes flashing past are sometimes repeated. Over time, a normal subject will become adapted to those. The neurons that encode them should become less responsive.

Turk-Browne and Kim will monitor her response by watching the fMRI for a drop in blood flow to a part of the brain called the scene-selective cortex, signaling that its neurons are becoming less active. The scientists are also looking at something else: in normal subjects, the brain will identify a scene as familiar even if there's been a time lag since the first exposure, and even if the subjects have been distracted by something else in between—playing an alphabet game, for example. If Lonni Sue does show adaptation, they want to know if it happens even when there's a time lag and a distraction.

Lonni Sue is all settled in now. The lights in the fMRI room have been dimmed. Nayeon Kim sits by her side, wearing a set of headphones so she can hear the scientists in the control room on the other side of the wall. In the control room, Turk-Browne and Jiye Kim sit at a desk with seven or eight screens arrayed in front of them. Aline stands behind them, watching closely. The screens show, among other things, a view of the fMRI scanner with Lonni Sue inside; a close-up of her right eye (both to monitor her eye movements and to note whether she seems to be dozing off); a static MRI scan of her brain, as a structural baseline for the changing images that will follow; and a duplicate of the images that will flash on her own screen inside the machine.

By now, Lonni Sue has been through the routine a dozen times or more. Before the scans begin, the scientists explain what she'll be doing. "Lonni Sue, we're going to be showing you pictures of places today," Turk-Browne says. "I think you're going to like them." What she needs to do, he explains, is that when she sees something she can identify as an indoor place—such as a kitchen, a living room, a hotel lobby, an office—she's to press the button she has clutched in her hand. If it's an outdoor place—a street scene, a beach, a mountainscape, a meadow—she does nothing. "I press if it's indoors?" she asks. Yes, that's right. She'll be doing eight or ten run-throughs, Turk-Browne expects, and he'll repeat these instructions each time as though she's never heard them before. As far as anyone knows, she might as well not have.

The first round starts. Images flash into view one at a time, for a fraction of a second, then disappear. Lonni Sue manages to keep up. "I'm not doing very well, am I?" she says despondently, her voice rasping through the low-quality speaker on the desk.

"No, no, you're doing great, Lonni Sue," Turk-Browne insists. Her single eye on the monitoring screen is wide open. She's fully engaged.

After they've run through the first set of images, all of them novel, it's time for the distraction. "Do you want to play an alphabet game?" Turk-Browne asks. It's still one of Lonni Sue's favorite things to do, a full seven years after Amy Goldstein introduced her to word-search puzzles and the joys of the alphabet. "You start," Turk-Browne says.

"Always," she says from her perch inside the scanner.

"Becoming," he responds.

"Creative," says Lonni Sue, over-enunciating.

Aline leans over and whispers in my ear: "You'll notice that the words she chooses are often uplifting, like 'creative' and 'inspiring.' She savors the words as she says them."

"Dandelion," says Turk-Browne. Lonni Sue giggles at that one. They stick with flowers for a while. She's particularly delighted when he comes up with "hibiscus." Lonni Sue finishes with "zucchini," and then it's time for the next round of images, to see if she shows adaptation despite the alphabet game and the time lag since the first round.

"Are you getting anything out of this?" Lonni Sue asks.

"Yes, we're getting some great pictures of your brain."

He patiently gives her the instructions again, since she can't remember them from one trial to the next. Aline whispers again: "Listen to how he's talking to her. He's really connecting with her. That's so important. These are more than just wonderful scientists."

Lonni Sue does the task again, followed by another alphabet game. It's animal-themed this time. She giggles when Turk-Browne says "iguana," but she's baffled when he offers "manatee." She's never heard of it. "It's a big, big . . . I don't know what it is, exactly," Turk-Browne tries to explain. "Maybe a fish?" He's not kidding. "I think it's a mammal," Kim tells him, gently. "It might be a mammal," he says into the microphone. He evidently doesn't watch the National Geographic Channel much.

Finally, about an hour and a half after it began, this test session is over. It's just one of a series trying to get at the question of what the medial temporal lobe's role is in helping the visual system adapt to the world. Another test, Kastner said a couple of months later, involved not scenes but line drawings—stars, circles, three-leaf clovers, hearts, and other, less generic shapes (a jagged horizontal squiggle, a seven-pointed star, a droopy star that looked something like a starfish, a snowflakey kind of thing, an object that might have been the bow decorating a dress). Testing Lonni Sue only on indoor/outdoor scenes and nothing else wouldn't give the scientists a comprehensive enough picture to say anything meaningful.

The results of all the experiments were still being analyzed when Kastner and I met, but the neuroscientists were already beginning to see a trend. The hypothesis going in was that with most of her medial temporal lobe gone, her visual system wouldn't show adaptation, especially after a long lag. But that's not what seems to be happening. When the scientists show Lonni Sue an object, then show her the same thing a couple of minutes later, the fMRI reflects a weaker response the second time. The information that this isn't a new object has to be stored somewhere during that interval. "So where is it?" Kastner asks. "I mean it's got to be stored somewhere, and no one believes that it would be stored in the object-selective cortex. So what controls it? We don't know."

Whatever it is, Kastner speculates that it might be part of the system in the brain that allows us to experience the world continuously from moment to moment, not as a series of random, disconnected events. Lonni Sue's memory has been devastated in many ways, but, Kastner says, "she has this kind of continuity and you can feel it when you spend time with her. She's comfortable in where she is right now. Many other amnesics do not have that." Clive Wearing is a good example: where Lonni Sue seems to roll through life in a generally happy mood, Wearing was constantly terrified. "It always makes me wonder whether he, for instance, does not have this continuity." And remember, she said, the current round of tests on Lonni Sue are looking just at the way she processes her visual environment. There may be other systems that process other aspects of our experience. If you felt as though you'd just awakened in a totally unfamiliar place, you'd be disoriented, too—and if this kept happening over and over and over, the word "terrified" doesn't really seem adequate to describe it.

Lonni Sue's experience of the world as continuous rather than disjointed might explain why she can work on her puzzles for long periods of time, why she can play a piece of music from beginning to end. But why does she have it while Clive Wearing doesn't? "I suspect Clive probably has more severe and more extensive damage," Kastner said. He hasn't been scanned like Lonni Sue has, so that won't be clear until his brain is autopsied after death. "It could also have to do with the fact that Lonni Sue is a visual artist," she said. "Her visual system may have a different repertoire than that of a musician. I'm just speculating here, obviously, but those are all interesting things to just think about and wonder about."

This particular set of experiments is unlikely to revolutionize neuroscientists' understanding of how memory works. It's a small piece in a very large puzzle. Henry Molaison's surgery in 1953, and the years of testing that followed, gave scientists the broad outlines of a solution to that puzzle. His case made it clear that the hippocampus is essential to the consolidation of experience into long-term memory. His case also showed that the brain has more than one memory system, and that at least some of them can operate without a hippocampus. But research on a number of amnesia victims, including Lonni Sue, has begun to indicate that it's not nearly that simple. Episodic memory and semantic memory might not be entirely distinct from each other after all.

Neither might procedural and declarative memory—"knowing how" versus "knowing that," as Gilbert Ryle called them in 1949, four years before Henry Molaison went under the knife. The experiments where Lonni Sue described artistic techniques and aviation lore, along with the tests of her ability to sight-read new music, have reinforced the idea that the line separating procedural and declarative memory is extremely fuzzy. As Lonni Sue put her jewelry back on in the fMRI control room, Aline told Turk-Browne about another example, which none of the scientists had heard about yet. They were chatting informally, and Aline was telling him about Lonni Sue's passion for horses when she was young—how she drew them and read about them and rode them, and identified so strongly with them that she ran hurdles on the Princeton High School track team so she could jump like a horse in a steeplechase.

"Could she still ride a horse, do you think?" Turk-Browne asked, his neuroscientific antennae suddenly pricking up. "Well!" Aline answered, in a funny-you-should-say-that voice. "I've asked her, 'How do you mount a horse?' And she can describe it beautifully, in rich detail." She can talk about how you don't really hold on with your hands when you ride, but rather with your legs. She can describe what it's like going over a jump, how you use your heels and the reins to guide the horse. "Oh, she knows quite a lot," Aline said.

Aline's observations weren't scientifically rigorous, but they seemed consistent with what the experiments had shown. Painting a watercolor, flying a plane, playing the viola, and riding a horse are skills that take time to master. Eventually, they become largely instinctive and procedural. Yet Lonni Sue, with vast, gaping holes in her medial temporal lobes, not only can still practice these skills (although her horsemanship and her ability to fly a plane haven't been tested), but also, crucially, can talk about how to do them, in remarkable detail.

Exactly what that means is unclear. "At least," Turk-Browne said, "she behaves as if she remembers some things. I don't know what the answer is," he added. "I think the medial temporal lobe is important for this kind of learning, but it may not be strictly required. Maybe it helps learn more quickly, but you can still learn on your own more slowly without something like that. There is a whole range of things we'll continue to look at."

*Chapter 14*

# MAGGI'S MEMORIAL

Lini has just moved the music stand, and Lonni Sue is not happy. "Put it back exactly where it was," she says. Then, too impatient to wait, Lonni Sue picks up the stand herself and moves it a foot or so to the left, repeating under her breath, "Put it where it *was*." Lini reaches for the microphone. "Don't move that!" says Lonni Sue. "It gets in the way of my bow." Finally, when everything is exactly where she wants it, Lonni Sue puts the bow to her viola and warms up with scales and arpeggios, and then plays a few passages from a courante by Johann Sebastian Bach. The sound is exquisitely rich, the intonation generally excellent. After a few minutes, she's ready to sit down, but Lini urges her to warm up a bit more. The sound technician hasn't finished setting up his levels, and needs another few run-throughs. "I've already played it two times," Lonni Sue complains. "I really want to get a seat before it's too late. Where do I sit?"

The date is Saturday, June 13, 2015. The sisters are in the main sanctuary of the Unitarian Church of Princeton. Eddie Johnson's memorial service was held in this room in 1989. Now Maggi Johnson's memorial is about to begin. It's no wonder Lonni Sue is on edge. She's in an unfamiliar situation, and even small changes in her routine are difficult to adjust to. Beyond that, Maggi had been a solid, calming influence when the Johnson women had spent time together. Today that influence is gone. Lonni Sue understands that her mother has died, and she also understands that she'll be performing for the people who are still finding their seats before the service begins.

About six weeks earlier, Maggi was in the kitchen, getting herself breakfast, and Aline was walking in from the dining room. She saw her mother take an orange, then pause, then begin to slump. A blood clot had abruptly blocked one of the arteries feeding her brain, starving her neurons of oxygen. At the age of ninety-seven, Maggi had just suffered a massive stroke. Two weeks earlier, she'd given a talk to the Princeton Art Alliance. Aline had helped her with the PowerPoint, but Maggi needed no assistance with the talk itself. "She spoke for an hour," Aline said later. "My mother had the audience fascinated," she said. I myself had seen Maggi at the gym at around the same time, pumping small amounts of iron under the watchful eye of her personal trainer.

Lonni Sue had been all alone when the encephalitis virus began to take hold in her brain. Fortunately, Maggi wasn't alone when she had her stroke. Aline managed to catch her mother as she fell, ease her gently into bed, and call 911. Even so, there wasn't much the doctors could do. Maggi had a few days at most to live, they said, and Lini was grateful to hear it. It would have been torture for her mother to linger on for weeks or years, bedridden, paralyzed, barely able to speak, and who knows what else.

That night, sitting at Maggi's bedside, mother and daughter had what Aline would later call "a miraculous conversation." Maggi could still speak, but only a little, so while Lini communicated in words, her mother responded in hand gestures, smiles, raised eyebrows. Having lived together for so many years, they knew each other's scripts by heart.

They talked about all the important things—their own lives, and how much Ed had meant to both of them, and their love for each other. Of course they talked about Lonni Sue. "You know I'll keep encouraging her," Aline said. In response, Maggi pointed at her daughter and raised her eyebrows, and Aline understood. Sometimes, in her intense focus on her sister, Lini would forget about her own needs. Maggi would gently remind her to take some time for herself. That's what she meant now. Yes, Aline answered, she would take care of herself and try to keep things in perspective. Later, Aline would tell me that the moment was very peaceful and intense and dear.

Then Maggi struck the pillow with the edge of her hand, as if severing something. It wasn't the first time that day that she'd done it. Finally, Aline understood this as well. It was a subject they'd talked about many times after Ed's death two and a half decades earlier. "Do you want to die?" Lini asked. Yes, signaled Maggi emphatically. "Do you want to live?" No. Once she was confident that Aline got it, she stopped making the gesture.

After a couple of days, Maggi stopped accepting food and water. She didn't want to prolong the process of dying. Even though she'd live for only a few days longer, however, Maggi had to leave Princeton Hospital. There was no possible treatment, so it was against the rules to stay. Aline would have to figure something out. She'd moved Lonni Sue from one rehab to another in upstate New York, and she'd untangled her sister's complicated finances. But she'd had Maggi for help and moral support during those times. Now Maggi couldn't help.

Fortunately, Aline didn't have to navigate this challenge by herself. Like all of the scientists working with Lonni Sue, Sabine Kastner had come to think of the Johnsons as friends, not just as a test subject and her family. One Sunday afternoon a year or two earlier, Kastner had invited the Johnsons to her house for what she called a German coffee hour. (Kastner was born and grew up in Germany.) "It was a beautiful late spring Sunday," Kastner said, "and we just had a good time. My family was there. I think Nick [Turk-Browne] came as well, with his wife. It was not much different from other family gatherings that we have."

Aline had sent out an e-mail blast about Maggi's stroke to everyone who might want to know, and as soon as Kastner found out, she came over to the hospital to offer whatever support she could. A doctor as well as a neuroscientist, she could help Aline understand what was going on medically. As a friend, she sat quietly with mother and daughter, just to show her affection and support.

It took a couple of days of complex logistics, but Maggi was finally transferred to the same place where Lonni Sue lives—a place where she could be made as comfortable as possible for her last days on earth. The stroke happened on a Wednesday. The following Saturday afternoon, Maggie lay in her room, just down the hall and around the corner from Lonni Sue's. She was in a sleep from which she would never awaken. Aline stood on her mother's right side, grasping her hand, stroking her forehead, and whispering words of comfort. Kastner was also there. She leaned over Maggi from the left, holding her other hand. Classical music played in the background. Their bodies formed an arch over the sleeping woman, as though they were trying to protect her from any more suffering. They spoke softly to her, reassuring her that everything would be all right, that Lonni Sue would be well taken care of, that it was okay to go.

As they hovered, Lonni Sue sat at the foot of the bed, drawing. Aline had told her she could bring in the folding table from her own room, so she wouldn't have to stop working on her puzzles. Lonni Sue would occasionally look up from her work to check on what was happening, a look of concern on her face. Every so often, Lini would try to engage her. "Lonni Sue, would you like to come and talk to Mum?" she would ask. "Would you like to sing to her?" Lonni Sue seemed unsure. At one point, Lini tried to impress on her how serious the situation was, that Maggi would soon be gone from their lives. "I get it, I get it," Lonni Sue said, impatiently. Her watch and her schedule told her it was time for dinner. She was conflicted—she lived by her schedule, but she didn't want to leave. She asked her sister what she should do, but Lini was no help. "Whatever you want," she said. Lonni Sue decided she had to go to dinner. Five minutes later, she was back. Her need to stay was stronger than her need for routine. During Maggi's last day, Lonni Sue would read to her mother from the catalogs of shows Maggi had taken part in. It was her own idea. When Maggi finally passed away, late on Saturday night, Lonni Sue told her sister, "It's a new chapter in Mum's life," and as Lini was leaving, exhausted, Lonni Sue told her, "Make sure you take care driving home, because it's just the two of us now." She really did get it—and, somewhat surprisingly, she held on to it. Lonni Sue wouldn't bring it up unless something reminded her, but when that happened, even many months after Maggi's death, Lonni Sue knew that her mother was gone. There were no echoes of the "Daddy died" conversation, probably because Maggi's passing was brought to her attention so frequently in the days just before and the weeks just after it happened.

For the service, Aline has drawn speakers from many different parts of her mother's life. Marie Sturken and Judith Brodsky, two Princeton-based artists who have known and collaborated with Maggi since the 1970s, represent the art world. Dudley Carlson, who had come with her husband, Curt, to those unforgettable chamber-music evenings four decades earlier, is there to share her experience of the Johnsons in the days before the girls departed to pursue their own lives. Barbara Landau steps to the podium to talk about what she called Maggi's last great project—helping bring Lonni Sue back from the brink and turning her tragedy into something positive.

Cover image, *The New Yorker.*

Cover image, *The New Yorker.*

TOP: Illustration for a *New York Times* column. *Copyright July 9, 2007, by Lonni Sue Johnson. All rights reserved.*

BOTTOM: Illustration for a *New York Times* column. *Copyright August 6, 2007, by Lonni Sue Johnson. All rights reserved.*

TOP: After Lonni Sue spoke one of her first coherent sentences ("Bird on a tree") six weeks after her illness, Aline illustrated it (top) and had her sister draw a copy. It was an early indication that Lonni Sue's signature style was returning. 

BOTTOM: In the fourth month of Lonni Sue's recovery, her mother played a drawing game with her: Maggi drew a line in red, then Lonni Sue responded in blue. They alternated until an image began to appear. Just as she had when her daughters were young, Maggi was once again serving as both mother and art teacher. She worked tirelessly, trying to get Lonni Sue back to her art. 

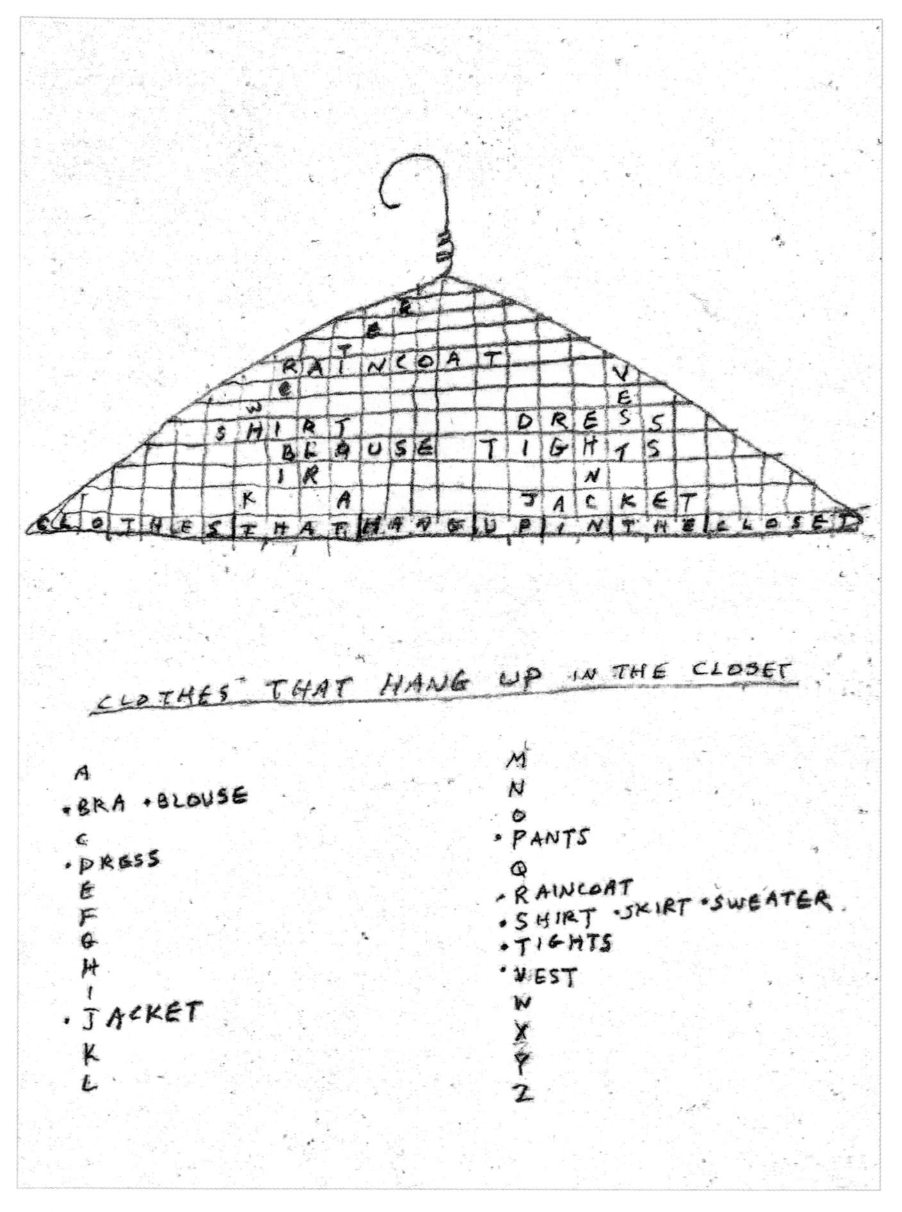


This is one of Lonni Sue's early illustrated word search puzzles; the drawing is relatively simple, but the pairing of image with puzzle theme is quite sophisticated, given her brain injury. 

As Lonni Sue's recovery art continued to evolve, the puzzles became even more complex. *Copyright March 27, 2009, by Lonni Sue Johnson. All rights reserved.*

This puzzle consists of five pages; only the first appears here. The other four pages are word grids. *Copyright April 13, 2009, by Lonni Sue Johnson. All rights reserved.*

**Lonni Sue Johnson**

**Plays the viola • Forever horse-back riding, reading, skiing, and traveling • Would like to meet Charlemagne • String quartets at PHS • "Quality in quantity."**

**French Club 1; Hockey 2; Ivy 3,4; Orchestra 1,2,3,4; String Orchestra 1,2,3,4; Track 1,2,4, Co-manager 3.**

TOP: Lonni Sue's high school yearbook photo, Princeton High School Class of 1968.

MIDDLE: Lonni Sue in her room with her viola, 2014. *Photo by Eileen Hohmuth-Lemonick.*

BOTTOM: Lonni Sue showing Maggi a drawing, 2014. *Photo by Eileen Hohmuth-Lemonick.*

Lonni Sue herself gets up three times—once to play the Bach piece on the viola; once to read the words she'd written about her mother in an apron, straddling the threshold between art and motherhood; and once to sing an alphabet song.

As the sanctuary gradually fills, people come to the front to say hello to Lonni Sue. A moment earlier, she was barking at her sister. Now she's basking in the attention of those who appear to be complete strangers, many of whom used to be her friends. "Who are you?" she asks Robert Landau. He introduces himself, kissing her once on each cheek. "Oh, you have beautiful lips," she tells him. "You must say that to everybody," he answers. Dudley Carlson comes up next. "Hi, Lonni Sue!" she says. "Hi," says Lonni Sue, trying to be gracious to this woman she doesn't recognize at all. Mike McCloskey sits down next to her. "Have you ever seen the books I wrote and illustrated?" she asks him. "Well, some of them I just illustrated." He'd love to see some, and she just happens to have several along with her, stuffed into her blue tote bag along with colored pencils and pens and folders full of paper. She's feeling too exuberant now to sit still, so she stands and pirouettes, her pink shirt flashing from under her black fleece vest. She mimes playing the viola. She sits again.

"There were some tense moments," Aline admitted afterward. Lonni Sue's irritation about the music stand was only part of it. Aline knew that her sister could never just sit patiently through a service for an hour or more. She set up a couple of TV tray tables in the front row so that Lonni Sue could work on her puzzles. She wrote out a detailed schedule for her sister of what was happening when, and in what order, which she'd shown Lonni a few days before the service. Big mistake. Partway down the page it said: "Lonni Sue sings a song." That had set her off. She insisted she had to write the song out beforehand, but Aline didn't like that idea. "It's when you improvise that the magic happens," she insisted.

In the end, Aline managed to distract her sister, and didn't show her the program again until just before the service began. But that was also a big mistake. "I very carefully planned how she would be cued," Aline said later, "so she would be fine when the time came. It was fine with me, but it was not fine with her, as I discovered." Again, Lonni Sue went into a mild panic. "When am I supposed to be onstage, and what will I do, and how will I know?" she asked. Aline showed her where her parts came on the program. "You can just bracket these, and write 'stage,'" she suggested. Lonni Sue bristled. "Don't tell me what to do or what to write!" she said. At this very last moment, with the service about to begin, Aline now had to go through the program with Lonni Sue, line by line. "I realized that my way of explaining to her, which would be fast and efficient, was not the way she needed to hear it," Aline said. "She had to ask her questions in her order and not rush, and then she would write it down." Aline's patience frequently seems to be nearly infinite.

Finally, it was time to begin. The sisters would start onstage, with Aline welcoming the guests. Lonni Sue wasn't happy onstage, either. "There's no light up here! How can I draw? This is terrible. It's very bad," she said. But Aline calmed her down, and stepped to the podium to welcome the guests. "Let us greet one another," she said, "the people in front of you, behind you, and on the side." She stepped away to greet Lonni Sue with a hug. Lonni Sue broke out in laughter. "Well put!" she said, delighted. "Well put, well put, well put, well put!"

Aline began by talking about both of her parents, and about their time in Japan. "In Tokyo, on Sundays," she said, "they would go on a walk in the park around the emperor's palace, and then they would go find a sandwich at the top floor of a building with large windows, and as they looked down on the view, they would dream together." She spoke about Maggi's artistic mentorship of Lonni Sue, both as a child and again as she recovered from encephalitis. Then she introduced Lonni Sue to do a reading. "Here I come," said her sister into the microphone. "I hope you can hear this all right, and I hope it's okay." It was perfect. She read with genuine feeling, rolling the words over her tongue as though each were a delicious morsel. When she came to the line "She liked to make prints more than cook," Lonni Sue started laughing so hard she couldn't continue. That was perfect too: the crowd erupted in laughter with her. When she was finished, Aline asked if there was anything else she wanted to add. "She'll always be alive for us all, because her memories will thrive. Amen. Thanks for listening."

A few minutes later, after Aline had spoken more about her mother's life, it was time for the alphabet song. Lonni Sue put down her drawings, stepped once again to the podium, and began to sing.

"Artist beautifully create delights . . . enthralling friendships growing happiness . . . inspirrrring . . ." She rolled the "r" theatrically, then broke up with laughter for a few seconds at her own silliness. "Joyfully known, lovely music nicely originating . . . pleasing quest relationships spectacularly treasured uniquely vocally wonderfully Xeroxing . . ." She broke up here, too, and so did the crowd. "Yearning zestfully." Throughout, she improvised a melody that wandered up and down the musical scale, lingering on some words, sailing through others. She lingered longest on the last word, embellishing it the way an opera singer would. When the crowd burst into applause, she blew them an enormous kiss. Then Aline led her back down from the stage to listen to the other speakers—which she did with only part of her attention, glancing up every now and again from the puzzles she continued to work on furiously.

Finally, it was time to play. "I hope you love music," she said to the audience, and then launched into the courante. It went off without a hitch. Virtually everyone in attendance knew what had happened to her, but few realized until now that someone with such profound amnesia could still read music and play an instrument so beautifully. "Thanks for listening," she said when she finished, beaming, as she stepped off the stage once again. On the way down, she passed Barbara Landau, who was making her way up. "Oh, hi!" Lonni Sue said in a loud voice. It was totally inappropriate, but utterly charming. When Barbara stepped to the microphone, she talked about the curious ways that her life had intersected the lives of the

Johnsons—an uncanny alignment, she called it. She and Maggi had estimated that they'd probably first met when Barbara was two years old, she said, when her family had lived just around the bend from the Johnsons' house. Her father had worked at RCA, just like Eddie. She and Lonni Sue both went to the Littlebrook School the year it first opened, in 1960, although Barbara was several classes ahead. She didn't even have time to mention her other important connection to Lonni Sue—her husband, Robert.

And then, after Dudley Carlson spoke of those wonderful chamber-music sessions, it was over. People came up to Lonni Sue, who was holding court at her folding tables in the front row, to say how much they loved the service, and especially her reading and playing and singing. "Do you like to draw?" she asked Joe Yacinski, her art-director friend from the New York days who had come up from Washington for the occasion. "Tell me your name again," she asked a woman who told her they'd known each other forever. "Anne Reeves," said the founding director of the Arts Council of Princeton. "You taught my daughter Emily," she said. "Oh, Emily Reeves!" said Lonni Sue, as though she remembered the daughter fondly. She didn't, but she played along so enthusiastically that you'd think she'd simply misplaced the name as we all do sometimes, and nothing more. "What was your name again?" she asked. "Anne Reeves." Oh, yes, of course!

Another gray-haired woman came up. It was Danae Meray-Horvath, whose unusual name I've always remembered from high school, just like I remembered Lonni Sue's, although I'd never spoken to either of them at the time. She has been to see Lonni Sue many times since the amnesia, so she understands that her old friend and fellow musician in the Princeton High School Orchestra can't remember much about her. "Lonni Sue and I have been friends since we were ten years old, and that was fifty-five years ago," she told an onlooker. "Oh, I can't believe how old we are!" Lonni Sue said—just the response you'd expect from someone without amnesia. Her memory is gone, but her social skills are perfectly intact.

Her mind was clearly troubled, though. As Lini said good-bye to the last remaining guests outside the church, Lonni Sue stood off to one side, talking with Landau. "I like to spend all my time on my work," she said, "and it's just wrong that she's not . . . I'm just not free. It's bizarre, what's going on." She brings this up every so often. She's not free, the way she used to be. She isn't allowed to fly. She can't leave the place where she lives, unless she has an escort. Since she's mostly unaware of her memory loss, it's hard for her to understand why her life is so limited. And now that she's found her new life's work—her puzzles—she's not free to do them every minute of the day. She doesn't want to be standing here, waiting for Lini. She doesn't understand why they make her go to meals, and make her go to bed at night, when she'd rather keep working feverishly. Landau tried to distract her. "Oh, my gosh, I didn't see your pocket full of pens," he said. "Two pockets," she corrected him. "You're fully loaded," he said. "But really it's just bizarre, what's going on," she said once more.

*Chapter 15*

# THE OPPOSITE OF AMNESIA

When Henry Molaison's hippocampus was removed in the 1950s, the catastrophic memory loss that followed took neuroscientists completely by surprise. In the year 2000, another patient turned up whose case was just as surprising, but in the opposite way. "I was minding my own business," James McGaugh, a neuroscientist at the University of California, Irvine, recalls, "and an e-mail showed up on my computer. A young woman said that she had a memory problem, and would like to meet me to talk about it." When he heard "memory problem," McGaugh naturally assumed that she couldn't remember things. No, she explained in a follow-up message, it wasn't that at all. Her problem was that she couldn't forget. The woman, whose name is

Jill Price, was thirty-four years old, and she claimed to remember everything she'd done or learned since she was eleven years old. Give her a date after 1974, and she'd tell you what day of the week it fell on, what she was doing that day, and what was going on in the world.

"I was very skeptical about this, naturally," McGaugh said, but he agreed to give Price some time on a Saturday morning. He didn't really know what to do with her. He studied memory systems, but in animals, not humans, and as far as anyone knows, animals don't have the kinds of explicit, declarative, autobiographical memories that humans have. You can teach a rat to remember a maze, but you can't get him to talk about what he was doing the other day. Since this all happened in 2000, there were a lot of books and magazine articles around at the time that looked back over the twentieth century. McGaugh had one of these books in his office; it reproduced year by year the most important newspaper stories from the entire century. "Someone had given it to me as a gift," he said. "I'd hardly even looked at it." When Price showed up, he pulled the book off the shelf and began leafing through it. He gave her a date and asked her what had happened in the news on that day. "She was absolutely errorless," he said.

Price was also more or less perfect in recalling events from her own life as well. McGaugh and his colleagues were able to corroborate this by talking to other family members, or by looking in Price's extensive diaries—fortunately, since without independent corroboration, there's no way of knowing whether someone is remembering an autobiographical event accurately or not. Price's memories are inevitably tied to a particular date, but according to McGaugh, it's a different sort of recall than what's involved in autistic-savant syndrome, in which someone with autism can tell you the day of the week for any date in history. "People like Jill Price can't tell me what days go with dates before they were born, or with dates in the future. What they can tell me is the dates of days they've personally experienced."

For several years, Price, who initially went by the pseudonym A.J. in published papers, was the only known example of what McGaugh and his colleagues called *highly superior autobiographical memory* (HSAM). Reports on her case generated some publicity, however, which convinced several others with the same condition to come forward. Price's autobiography, *The Woman Who Can't Forget,* published under her own name, brought in even more subjects, and a report aired by *60 Minutes* in 2010 was what McGaugh calls "the big hit. It generated literally hundreds of people who contacted me to say that they had this ability." Not all of them truly did, but McGaugh ended up with about fifty of what he calls "bona fide subjects," who have now been studied in some detail. Unlike Price, who found the load of memories she carried to be a burden, many of the subjects, including the actress Marilu Henner (who starred on the sitcom *Taxi*) and the TV producer Bob Petrella (*Ice Road Truckers*), think it's fun.

Ideally, the neuroscientists who study people with HSAM are hoping to find insights into normal memory by looking at these subjects' exceptional memories—the mirror image, in a sense, of what researchers have been doing with amnesics ever since Brenda Milner began working with Henry Molaison in the 1950s. With Molaison and Johnson and the other amnesia victims, the hope is to look at what's missing in their brains and figure out how the loss affects memory. With HSAM subjects, the idea is to see in what ways their memories are supercharged and try to understand what might be different in the physical structures of their brains.

So far, says McGaugh, the research hasn't produced any deep insights, but the scientists have come to a few broad conclusions. One is that the HSAM group isn't any better at learning new information than anyone else. They're just worse at forgetting. That doesn't apply to every detail of their lives; their memories aren't photographic. But some of the memories they hold on to are clearly not burned into their consciousness based on how important they are. They're likely to remember things like the weather on a particular day—clear evidence that they didn't make a special effort to store the information, since why would you? Their memory feats evidently don't require any special effort at all, in fact. They just remember things.

That puts them in an entirely different class from memory champions, who enter contests where they have to memorize hundreds of random digits, or the exact order of cards in five decks shuffled together, or some other useless information. It turns out that anyone can do that, as the journalist Joshua Foer showed in his 2011 bestseller *Moonwalking with Einstein*. The book chronicles Foer's attempt to compete in the USA Memory Championship. He ended up winning, but like all memory champions, he did it with an age-old trick, known at least since the time of the ancient Greeks. It's called the method of loci, or the memory palace: you simply imagine yourself traveling through a familiar landscape or building, and tag each thing in the list you're trying to memorize with a particular location along the journey. Or instead of locations, you can use imaginary scenes so weird you can't forget them. The title of Foer's book reflects one of these. "If you try to picture Albert Einstein sliding backwards across a dance floor wearing penny loafers and a diamond glove," Foer writes on his Web page, "that's pretty much unforgettable."

The technique is great for remembering lists, or stanzas in an epic poem, or the orders of cards, but it doesn't have much to do with memory as we ordinarily think of it. Although he became the U.S. memory champion, Foer admits that he still forgets where he put his keys—as does Jill Price.

McGaugh and other researchers have also discovered that HSAM subjects tend to rank higher than average on tests of obsessive-compulsive tendencies, which might be a clue to what's going on inside their heads. If they tend to ruminate more than average about what happens to them during their lives, they might burn that information into their memories with special power.

I once tried doing this deliberately. I was hitchhiking through

Europe with Jim, my manual-transmission instructor, during college, and we ended up in Communist East Berlin, in what was then still a divided Germany. (We didn't sneak in; it was legal for Americans to visit for a few hours.) At one point I was on a curbside eating a bratwurst, looking across the low-rise skyline at a tall concrete TV tower with an observation deck at the top—the only modern structure in that half of the city, as far as I could tell, where buildings half destroyed in World War II bombings still hadn't been repaired. For whatever reason, I decided I wanted to remember that moment. I looked hard at the tower, felt the curbstone beneath my hand, felt the unusual warmth of that April day on my skin—and sure enough, I remember it vividly, forty years after the fact. The other moments I remember from that trip—seeing Pope Paul VI celebrating Mass at the Vatican, telling a soldier in Yugoslavia that no, we had no drugs to sell him, standing up on a train at three in the morning with the worst heartburn of my life (I'd eaten a lump of fried dough on the platform before boarding)—all of these were memorable in and of themselves. Not this one.

The same might be true of the HSAMs. They might be rehearsing events, obsessively if perhaps unconsciously, making them impossible to forget. Evidence for the obsessivity connection also comes from MRI scans of their brains. A region within the structure known as the *striatum* appears to be larger in these subjects than normal; the same region has also been implicated in obsessive-compulsive disorder. Additionally, HSAMs seem to have a more active uncinate fasciculus, which is a pathway that carries signals between the frontal and the temporal lobes. "This finding is intriguing," wrote McGaugh and his colleague Aurora LePort in a 2014 article for *Scientific American,* "because of evidence that injury to this pathway impairs autobiographical memory."

All of this might point to something important, although the researchers looking at HSAMs aren't sure what it might be. "This is like arriving on the scene of the accident, and you're trying to figure out who caused what and all the rest," McGaugh said. "For my entire life, my scientific life, I've been an experimentalist," he said, "and so I would do an experiment. This group gets this treatment, that group gets that treatment, and then you look at the difference and draw conclusions and write a paper." This is the reverse of that, he said, echoing what Mike McCloskey had said about Lonni Sue. "The phenomenon appears, and you try to work backward and say what the hell caused this, and it's just more complicated."

It's not even clear that super memory necessarily first appears at about the age of adolescence, although this is when most HSAMs begin to notice it. Maybe it's been there all along. "We have a couple of nine-, ten-year-olds that seem to have it, too," he said. The reason most of the current subjects believe their extraordinary memories came along later, he and his colleagues suspect, is that puberty is the time most of them discovered that everyone doesn't share this ability. McGaugh and others are just beginning to look at younger kids to see whether their gut feelings are correct.

*Chapter 16*

# OTHER AMNESICS

A man wakes up in a cheap motel room. He doesn't recognize the place. He doesn't know how he got here. He doesn't know how long he's been here. He rummages through the drawers—nothing. He looks in the mirror. His body is covered with tattoos that say things like "Find him and kill him" and "She is gone. Time still passes" and "John G. raped and killed my wife" and "Never answer the phone." He finds a handful of Polaroid photos, mostly of people, with cryptic notes scrawled on them, including one that says "He's Teddy. Don't believe his lies. He's the one. Kill him."

That's how the movie *Memento,* which was released in 2000, begins. Before long, you learn that Leonard, the man in the motel room, has anterograde amnesia. Like Lonni Sue Johnson and Henry Molaison, he can't form new memories. But in Leonard's case, he remembers everything that happened before his brain injury with crystal clarity. You learn that Leonard's every waking hour is devoted to a mission. He's determined to take revenge on the man who raped and murdered his wife before cracking him on the skull and causing what he calls "my condition." Since he can't remember anything that happened after his brain injury, he takes photos, scribbles notes, and has himself tattooed with essential information that will help him recall any information he learns about the killer.

As is almost always the case with a movie, some of the scientific details are dubious. In the real world, victims of anterograde amnesia almost always suffer from some degree of retrograde amnesia, the inability to bring up memories from before the injury, as well. The loss tends to be far more complete for autobiographical memory than for semantic memory, and it's most acute for the years preceding the onset of amnesia. Just as with Alzheimer's patients, the distant past is clearer than the recent past. Yet even though the attack happened literally seconds before he got the brain injury that robbed him of his memory, Leonard recalls not just the semantic fact that his wife was murdered but also the specific autobiographical details of how it unfolded. Not only that: Leonard is also fully aware that he has anterograde amnesia. He keeps telling people about what he calls "my condition," although he keeps forgetting that he told them. (He also insists that his condition isn't amnesia, which director Christopher Nolan evidently believes applies only to forgetting the past.) Real amnesia victims don't have that insight. In the early years, Lonni Sue might concede that she had a minor memory problem. Now she doesn't admit to any problem at all. "I think my memory is pretty good," Lonni Sue has told me more than once. Aline can't really say when that transition happened.

Still, the representation of amnesia isn't bad for Hollywood. Unlike, say, 50 *First Dates,* which clinical neuropsychologist Sallie Baxendale, of University College London, has written "bears no relation to any known neurological or psychiatric condition," *Memento* at least gets the sense of what amnesia must feel like at least partly right. It does so through the trick of putting scenes in reverse chronological order. When you see something unfolding on the screen, you have no idea what came before. "What they captured extraordinarily well," Neal Cohen said when I visited him at the University of Illinois, "is the sense of a little island isolated from the rest of time, in which all is unconnected to the next thing or the thing that preceded it." Princeton's Ken Norman assigns this film as required viewing in his course along with *Eternal Sunshine,* as many other professors of neuroscience undoubtedly do as well.

The fictional Leonard does things that would be impossible for a real amnesic, but that doesn't make him entirely unrealistic. At least one amnesic in the literature also does things no amnesia victim should be able to do. Neal Cohen, who studied the woman known publicly only by the pseudonym "Angie," told me about her when I visited him. Angie was twenty-nine, a special-ed teacher, working on her Ph.D. in education, when she had a bad reaction to a flu shot and went into anaphylactic shock. Unfortunately, she was driving at the time. She slammed into a telephone pole and sustained a major head injury. "It was not a great outcome," Cohen said. Angie's medial temporal lobes were badly damaged in the accident, just as Henry Molaison's were in surgery and Lonni Sue Johnson's were by encephalitis. She came out of the hospital with severe anterograde amnesia. This was in 1985.

What happened next was more awful than it had to be, according to a paper Cohen and several colleagues published in the *Journal of Clinical and Experimental Psychology*. Angie's parents had her discharged from the hospital and set her up with an apartment and a dog—an untrained dog, according to the researchers. She had no access to treatment. Her friends rallied to help her for a while, but eventually they drifted away. So did her fiancé. She tried to go back to work after about a year; her profound memory problems made that impossible, however.

"It's a terrible story," Cohen said. "Except she was a remarkable person. She *is* a remarkable person." Bit by bit, Angie built herself a new life. She moved to a new town. She went back to graduate school. She met someone, and got married, and helped raise the man's three children from a previous marriage. She got a job as a project manager for an educational testing company, and did the job successfully. She developed several new, close friendships with women she would go out and even travel with. "You say, 'Oh, come on. How is that possible?'" Cohen said. "But if you go to her house she looks like any other overburdened soccer mom. Little stickies everywhere, notes everywhere." She uses the same tools you or I would use when we have a million things going on. "And then, she's extraordinarily good at structuring her life to be as orderly as it could possibly be." Angie learned, for example, through relentless practice, what the driving route was from home to the supermarket and back, and from home to the kids' schools. "But she couldn't concatenate [that is, link] them," Cohen said. She couldn't drive from the supermarket to the school. And if there was a detour somewhere along one of the familiar routes, Cohen said, "forget it. She just had to drive back home."

At work, Angie was equally organized. If you asked her what she did for a living, she'd answer, "I manage projects." If you asked her what kind of projects, she would have no idea. She couldn't tell you what the job entailed, or the names or any other facts about the people she worked with. Nearly any facts, anyway. She did tell the scientists that she had to watch out for one of her coworkers because "she'll stab you in the back," but couldn't say how she'd come to feel that way. She didn't know what her salary was. Again, said Cohen, "you're thinking, 'How is that possible?' The answer is that she was good. She was really, really good." She was a master at creating systems that served as external memories to replace the functions she'd lost—extensive notes recording her thoughts, color-coded folders, a heavily annotated calendar, all with multiple levels of redundancy. In order to have as normal a life as possible despite her severe limitations, Angie was determined to come across that way to others. There was no way she was going to let her coworkers know how profound her impairment was. In staff meetings, she would have her subordinates report, one by one, on the progress of their projects.

While they talked, she took notes and then reiterated their statements. For her, these were overt strategies for compensating for her memory impairment. For her employees, these actions were interpreted as an open leadership style in which their boss provided validation of their ideas and suggestions.

You can almost see the scientists who wrote the report shaking their heads in disbelieving admiration as they describe her.

To put it bluntly, Angie is a gifted con artist, albeit with a much more honorable motive. She's acutely aware of her limitations, and uses her sharp intellect and unusually high social intelligence to make her seem as normal as possible—and she pulls it off. In the course of their research, members of Cohen's lab talked with her about the unrelenting challenge of keeping up appearances. It was clear that she found it exhausting. She ended up taking early retirement due to the strain. She also stopped participating in the neuroscience research.

Lonni Sue is more densely amnesic than Angie, but she has also latched onto systems that give her life some sort of structure. One of these is her puzzle-making—the self-assigned job at which she works with relentless focus, enthusiasm, and joy. She didn't discover word puzzles for herself, but she has made them entirely her own. Another is her schedule. The place where she lives puts out a printed schedule every day, with information about mealtimes and activities and events. In the early days, this was meaningless to Lonni Sue. She'd simply write and draw all over it. Each day's schedule would wind up as part of what Aline calls a "landslide of papers," piled on one side of her desk. Once she became immersed in her puzzle world, however, Lonni Sue needed to know when she was likely to be interrupted. Although she resented it, she understood that she needed to eat, and exercise, and see doctors every so often. But she wanted to know when these things would happen and how long they would last, so she began paying close attention to the schedule. Nowadays, it's either on her desk or in the blue cloth bag she carries with her everywhere. It's always sticking out of the bag, Aline told me, so Lonni Sue can grab it and refer to it whenever she wants. She does that several times every hour, checking off things as she completes them and annotating it with new items—research sessions at Maggi's house, visits from old friends, appointments with doctors outside the facility—as she learns about them. With no memory of what she did yesterday and no sense of what she'll do tomorrow, the daily schedules give her an artificial timeline.

This lack of a time sense appears to be common to many people with profound amnesia, but it wasn't something neuroscientists thought a great deal about until Kent Cochrane came along. Cochrane, a Canadian, suffered a severe head injury in 1981, at the age of thirty, in a motorcycle accident. It wasn't his first. At sixteen, he had been knocked unconscious by a bale of hay that fell on him while he was working on his aunt's farm near Montreal. A few years later, he broke his jaw when his homemade dune buggy crashed into a truck. He didn't lose consciousness that time, but the next accident—he flew off the road as he was driving his motorcycle down a highway exit ramp—knocked him out for three days. He was "stuporous" for the next five and, as in Lonni Sue's case, appeared to recognize his mother only at about day seven.

When he finally came to full consciousness, it was clear that Cochrane had lost his ability to form new memories, just like Henry had, and that his episodic memories of the past had been almost entirely destroyed. A few years after his accident, Cochrane came to the attention of memory researchers purely by chance. "One of my undergraduate students came into my class one day," Morris Moscovitch, a neuroscientist at the University of Toronto, said, "and said, 'I think I've found another H.M.'" The student had come across Cochrane while volunteering at a sheltered workshop, a place where people with disabilities were given simple jobs to keep them occupied. "We brought him into the lab and tested him, and indeed he had a very severe amnesia."

What made Cochrane's case especially important, Moscovitch said, was not just the fact that he had such a distinct mix of lost and retained functions—his memory for past events was gone, along with his ability (mostly, anyway) to form new memories, while his semantic memory was largely intact, and he was otherwise cognitively normal. "It was that he showed up at around the time when people were starting to think about different memory systems."

That's not literally true, of course. As early as 1890, the Harvard psychologist William James had talked about the difference between short-term memories, of the past couple of minutes, and memories that go into long-term storage. In 1949, Gilbert Ryle had distinguished between "knowing how" and "knowing that"—implicit (or procedural) versus declarative memory. The distinction between episodic and semantic memory, however, which in retrospect was clear from the time of Brenda Milner's first experiments with H.M., had taken neuroscientists a couple of decades to work out.

One of the scientists who helped identify the difference was Endel Tulving, who had been born in Estonia but spent most of his professional career at the University of Toronto. In fact, Tulving invented the term *episodic memory,* which he described in a book titled *Organization of Memory,* in 1972. By the time Tulving began to study Cochrane a decade later, the idea that episodic memories were different from semantic autobiographical memories was only just taking hold among neuroscientists.

It wasn't clear, however, why such a distinction should exist in the first place. Episodic memory is great for telling stories about things that happened to you long ago. Many of the basic memories we rely on to get through life aren't episodic, however; they're semantic memories, of how to walk and talk and get dressed and recognize people we love and get from here to there. Clive Wearing's episodic memory was devastated by encephalitis, but he could still recognize his wife and remember how much he loved her. He could still conduct a choir and play the piano. Similarly, Lonni Sue can still read, write, play word games, play viola, and draw like a bandit. The fact that she can't remember her past except in generalities is a tragedy of sorts, but how essential is it, really? The fact that she can't form new memories is the more acute problem.

Tulving noticed something else about Kent Cochrane, however, which might be a clue to why explicit memory is so crucial. "One revealing thing that Tulving pointed out," Moscovitch said, "is that loss of episodic memory not only affected Cochrane's memory of the past but also impaired his ability to imagine the future. He couldn't describe what he was going to do in a week's time in any more detail than he could describe what he did a week ago." The same had been true of Henry Molaison, and it would turn out to be true of Lonni Sue as well. If you asked Henry what he was going to do the next day, he would invariably answer, "Whatever's beneficial." He didn't really have any idea. "People knew about that," Moscovitch said, "but we didn't play it up as much as we should have and neither did Sue Corkin or Brenda Milner."

But Tulving thought this was a big deal. "Unidirectionality of time," he wrote in a 2002 article in *Annual Review of Psychology,*

> is one of nature's most fundamental laws. . . . Galaxies and stars are born and they die, living creatures are young before they grow old, causes always precede effects, there is no return to yesterday, and so on and on. Time's flow is irreversible.

The only exception to this inexorable flow from past to future, he wrote, is human memory—in particular, the memory of specific events from our past. What distinguishes our episodic memory from our semantic memory, Tulving argued, was that the former is in essence a form of mental time travel. You don't just recall facts about yourself when you bring up an episodic memory; you re-immerse yourself in the past. For a short while, it's as though you're living there once again.

When Aline Johnson walked up to me on the streets of Princeton to tell me about her sister, it put me right back in middle school for just a few moments, complete with sights, sounds, and feelings. In order for that to happen, according to Tulving's theory about episodic memory, three things needed to be true. First, I had to have a sense of myself as existing in time, with a past, a present, and a future. There's no evidence that even the most sophisticated animals have this sense, Tulving asserted. "They have minds, they are conscious of their world, and they rely as much on learning and memory in acquiring the skills needed for survival as we do," he wrote. But their consciousness is confined to a sort of timeless state—a permanent present tense, as Suzanne Corkin would put it.

The second requirement for episodic memory, Tulving said, is something he called *autonoetic awareness,* the conscious awareness of our own mental state. When Aline's face transported me to middle school, I understood that I wasn't actually there—that the past is different from the present, and that this is where my mental images were coming from. People with schizophrenia who hear voices or have other disturbing hallucinations are said to suffer from *autonoetic agnosia,* a lack of awareness that what they're experiencing is a mental state, not reality. Finally, the third requirement, Tulving wrote, "is that mental time travel requires a traveler. No traveler, no travel." He was talking about a sense of self, the sense that there's an "I," a unique individual with a particular history. The primary reason Kent Cochrane had no episodic memories of the past, and couldn't form any new ones, Tulving concluded, was that he had lost his sense of existence in time.

The same appears to be true of Lonni Sue, to a large extent. She is no longer able to visit the past mentally; she can talk about it only in a general way, as though she's consulting an encyclopedia entry about her life. She also can't imagine the future. H.M.'s plan for tomorrow was always "Whatever's beneficial." Ask Lonni Sue what she'll be doing tomorrow and her answer is, "I have to check my schedule." She can't remember what happened more than a few minutes ago, and without a rich memory of the past she can't make predictions about what will happen in the future. But with that printed, annotated schedule, says Aline, "she's not floating around in time anymore." If you ask what she did this morning, she can consult her schedule and give you an answer. If you ask what she's doing later today, same thing. Without the schedule, none of this would be possible.

Losing a direct sense of yourself as having a past, present, and future doesn't affect semantic memory, because semantic memories are unanchored to a personal sense of time. When Henry Molaison told Suzanne Corkin, "We used to go on vacations along the Mohawk Trail," it wasn't attached to any particular episode. It's easy to mistake this for an autobiographical memory, not a semantic one, because we actually tend to talk about our past in generalities most of the time. "When you ask people about something they did in the past," Moscovitch said, "you don't really expect to hear a lot of detail. You want to know, 'Did you go on vacation to France?' 'Oh, yeah.' 'Did you have a good time?' 'Definitely!' 'Did you do anything interesting?' 'Oh, yeah. We went boating.'" That's good enough for most conversations. It's not clear unless you probe further whether this is a semantic or an episodic memory. The same is true for Lonni Sue. When she says "I used to fly an airplane" or "I drew covers for *The New Yorker,*" it sounds as though she might be thinking about specific incidents. It's only when you ask, "What was the subject of your first *New Yorker* cover? Where were you when you found it had been accepted? How did you feel?" that she'll start getting vague, and you'll realize that she can't come up with an answer.

Beyond that, most episodic memories tend to be converted into semantic memories with the passage of time. There really was a first time you heard that Paris is the capital of France, or that Franklin Roosevelt was president during World War II. For a short while, that might have been an episodic memory, but no more. "Just a week from now," Moscovitch said, "if somebody asks you about the conversation we're having right now, you'll be able to tell them the gist of what we talked about, not the details. Two years from now you'll maybe remember that you called me, but you won't have a clear idea of what we talked about except that it had to do with amnesia." He's right. Even Jill Price, the first person diagnosed with highly superior autobiographical memory, doesn't remember literally every moment from every day of her life. Not even close.

Early on, therefore, while it was impossible to avoid noticing that H.M. couldn't form new episodic memories, it wasn't nearly as obvious that he couldn't access old ones. Milner and Corkin actually thought he could, at least from his childhood. As a result, neuroscientists thought for years that the hippocampus and its neighboring structures in the medial temporal lobe were involved only in forming new memories, and that once these had been formed, they were permanently there. Once the host at the cocktail party introduced the guests to one another, his or her services were no longer needed. The introduction took a while, they thought; it might be several years before a memory was fully consolidated into long-term storage. But once that was done, so was the job of the hippocampus.

The work of Tulving and Moscovitch and many other researchers, working with both humans and animals, made it clear that this wasn't the case after all. Memories are stored throughout the brain, but for those special memories we can call up vividly, Moscovitch said, "the hippocampus is always involved in recovering and reconstructing the past." If the hippocampus is gone, or if its connection to the rest of the brain is severed, those rich episodic memories are lost. Moscovitch and Lynn Nadel, of the University of Arizona, made this point in a 1997 paper. "Suzanne Corkin went back to H.M. and looked at her notes and tests and so on," Moscovitch said, "and realized it's true: H.M. doesn't have any rich episodic memories of the past."

Being unable to call up episodic memories of the past might be embarrassing, but it's not in itself debilitating. Being unable to form new episodic memories is more problematic, but for someone like H.M. or Kent Cochrane or Lonni Sue Johnson, even that may not be the worst consequence of this particular form of brain damage.

It was Tulving's great insight that an inability to imagine the future might be the real issue. Patients with medial-temporal-lobe damage turn out to have terrible difficulty in solving problems in which you have to rely on explicit memories of the past—how to get a new job, for example, or making friends in a new neighborhood, or making up with a boyfriend or girlfriend after you've broken up were the examples Moscovitch offered.

Normal people can at least come up with a strategy for doing these things. But how well you do them depends on how richly you can create a mental simulation—how vividly you can imagine a solution, based on your memories of similar situations in the past, and the outcomes. "There's a high correlation between how rich your imagination is and how well you can solve these problems," he said. Since human existence is more or less an exercise in continual problem-solving, the inability to access your past and to imagine your future is truly catastrophic. This doesn't prevent Lonni Sue from living a contented, even joyful existence. But she can't devise strategies for solving any but the simplest of problems. She has difficulty making decisions, as her wavering between dinner and her dying mother made clear. There's no way she can function independently in the world.

More than a half century after Henry Molaison had his brain tissue suctioned out by an unsuspecting William Scoville, neuroscientists understand that the term "memory" covers many different brain systems that work together to allow us to navigate the world. The boundaries between them aren't always as clear as the textbooks suggest: Lonni Sue's ability to sight-read music on the viola, or to improvise an alphabet song, or to describe in detail how you prepare and paint a watercolor, aren't entirely declarative, but they aren't entirely procedural either. If Tulving is right, however, episodic memory is in a special class. It is, he wrote, an "evolutionary frill," a sort of icing, a layer that sits on top of the memory functions we share with other, less sophisticated animals. It looms large in our assumptions about what memory is, but it's probably a tiny fraction of the amount we've learned in total. "As a general matter, our obsession with the part of memory we have direct access to might be slightly misleading," says Bill Bialek, a biophysicist at Princeton who has studied the brain extensively. "I don't know. It might be that it's very special to us, and thus, really interesting, but we're missing the larger part." It was these other memory functions our prehuman ancestors needed for survival—procedural memory, so they could learn to use tools; statistical learning, so they could figure out where best to find food and how to avoid predators; and associative learning, so they could acquire and retain information about what plants were nutritious and what plants were poisonous. Episodic memory wouldn't have helped them very much in the struggle to live and reproduce. But modern human society would be impossible without it.

*Chapter 17*

# THE SAGA OF HENRY'S BRAIN

I pulled the *New York Times* out of its plastic delivery bag on the morning of December 5, 2008. It was about a year since Lonni Sue had begun to feel the first stirrings of viral encephalitis, although I wouldn't know this for many years. I unfolded the paper, and read this headline on the front page: "H.M., an Unforgettable Amnesiac, Dies at 82."

He was certainly unforgettable to me. I'd first read about H.M. in my freshman psychology textbook at college in the fall of 1971, less than twenty years after the surgery that robbed him of his memory. The idea of being unable to form new memories seemed appalling to me, and, along with the two hundred or so other students in the class, I tried to imagine what such an existence might be like. Naturally, I failed. I would come back to H.M.'s case several times as a science journalist, whenever I wrote a story that touched on the science of memory. I never knew his name, thanks to Brenda Milner's and Suzanne Corkin's absolute insistence that he remain anonymous while he was alive. But there it was at last, in paragraph seven: "On Tuesday evening at 5:05," it said, "Henry Gustav Molaison—known worldwide only as H.M., to protect his privacy—died of respiratory failure at a nursing home in Windsor Locks, Conn."

Nearly seven years later, I sat in Suzanne Corkin's office at MIT, talking with her about her long relationship with Henry, as both a scientist and as the closest thing he had to a relative after his mother died. I was asking about his death, and the postmortem research that she'd been preparing for in anticipation of his passing. "We've barely begun that research," she said, and then looked at me expectantly. I looked back, blankly. "Your next question," she prompted me, "should be 'Why?'" I still didn't get it. "Why . . . what?" I asked. "Why seven years later we do not know anything about the details of his brain?" Seven years did seem like a long time, now that she mentioned it. I assumed that this represented the slow and careful process of science.

I was wrong.

Immediately after Henry died, Corkin said, a meticulously detailed plan went into place to preserve the organ that had been surgically altered fifty-five years earlier. William Scoville's surgical notes, along with X-rays, CAT scans, and structural and functional MRIs, had delineated the broad outlines of the damage Scoville had done. He'd removed the front sections of both of Henry's hippocampi, along with much of his entorhinal, perirhinal, and hippocampal cortices, surrounding structures that relay signals between the hippocampus and the rest of the brain. Scoville had also taken out the amygdalae, which process the experience of emotion. The testing that Milner, Corkin, and dozens of others had done, meanwhile, revealed how this physical damage affected Henry's memory systems. Between them, these two kinds of research had given neuroscientists enormous insight into how the brain's physical structure allows us to remember the past and to create new memories.

Even the most powerful brain scanner, however, shows the structure of the brain with only a relatively coarse resolution. It can't tell the scientists precisely how much of the parahippocampal cortex remains, or exactly what's left of the entorhinal cortex, or how much the damage might extend into other brain structures. Surgery, encephalitis, and traumatic injury are all crude enough in their effects that each amnesia victim will have a different pattern of destruction. Each victim also comes into an amnesic state at a different age, with a different baseline of memory function before the injury (Nick Turk-Browne's brain, for example, works well enough that he's a highly regarded professor at Princeton, but he told me that he has a terrible memory for faces). Even the location of different functions within the brain, while similar from one person to the next, isn't exactly the same.

For all of these reasons, no one patient and no one case can tell neuroscientists how memory works. The more examples researchers have to work with, the better they can understand what's going on—and they need to know about both the structure and the functioning of each patient's brain in as much detail as they can, in order to get the fullest possible picture. It's relatively straightforward to test an amnesic patient's memory function, but the only way to study the physical structure of his or her brain in detail is to cut it open after death. Corkin is so fixated on the importance of doing this that she asked me several times during our conversation whether the Johnsons have made plans to have this done after Lonni Sue is gone (they hadn't, as of the time we talked).

In Henry's case, it was especially important to study his brain in detail because doctors didn't know whether his severe epilepsy had caused some brain damage that might have contributed to his memory problems. Over the last decade of his life, moreover, he had been going through a cognitive decline. "When he died," Corkin said, "he was very demented. He couldn't do anything for himself, and he was mute." Her best guess was that his dementia was vascular—that is, the result of small strokes, each of which had added a tiny bit of damage to the cognitive parts of the brain, on top of his memory loss. But maybe it was Alzheimer's. Or maybe it was Lewy body dementia, a less common degenerative disease of the brain, in which protein deposits affect not only memory but also motor control. (You can think of it as a cross between Alzheimer's and Parkinson's diseases; the actor and comedian Robin Williams reportedly suffered from Lewy body dementia toward the end of his life.) Or maybe it was several kinds of dementia at once. "When people are demented," Corkin said, "they don't necessarily just have Alzheimer's disease. They might have Alzheimer's and Parkinson's or Alzheimer's and Lewy body." Understanding which of Henry's symptoms came from which source could sharpen scientists' understanding not only of how his surgery had affected him, but also of what role dementia played.

For example, Alzheimer's and Parkinson's, scientists now believe, might start to affect the brain years, or even decades, before symptoms appear. In a groundbreaking study of 678 nuns, for example, researchers found that aging sisters who developed Alzheimer's had shown consistently less sophisticated writing skills as young women than their counterparts who didn't end up with the disease—specifically, their writing was lower in the expression of positive emotion, and also in what the researchers called "idea density" and "grammatical complexity." The data came from essays they'd all written shortly after entering the convent in their early twenties; since all of the young novices had to do the same writing exercise at roughly the same age, and because they'd come from very similar ethnic and economic backgrounds, it proved to be an invaluable resource for comparing the women's cognitive function at an early age. The scientists can't know for sure whether the poorer writing skills were the very earliest signs of dementia, or whether there is some reason, still unknown, that poor cognitive function is associated with a higher risk of Alzheimer's. The fact that this is even a possibility, however, means that it's crucial to understand Henry's brain damage in the finest possible detail, so that the effects of other brain disorders can be teased apart from the effects of his surgery. It's the crucial final stage in the research project Brenda Milner had begun more than fifty years earlier.

The brain is an extraordinarily fragile organ, however. It has a consistency like that of Jell-O, or of tofu, depending on whom you ask. That's why William Scoville was able to remove sections of Henry's medial temporal lobes, not by slicing, but essentially by sucking them out through a straw. In order to study Henry's brain postmortem, it had to be removed and preserved with extraordinary care—and very quickly, too, because it would start to deteriorate the moment Henry died.

For that reason, Corkin had begun planning the removal as early as 2002, and had ultimately come up with a strategy that had the precision and timing of a commando raid. "My colleagues and I had drawn up a flowchart of who needed to be contacted and in what order when Henry died," she writes in *Permanent Present Tense*. "My assistant had laminated a wallet-size version of the flowchart for each of us to carry with us, and I kept copies under the wall-mounted phone in my kitchen, in my car, in my office, and on the desktop of each of my three computers."

The call came just before five-thirty in the afternoon on December 2, 2008, she recalled, as she was sitting in her car, just after arriving home from work. Henry had been pronounced dead less than twenty minutes earlier. Her first act, which she did before she even got out of the car, was to call Jacopo Annese, a neuroscientist at the University of California, San Diego. Annese had written a grant proposal to the National Science Foundation a couple of years earlier—"a sort of manifesto for what I would do with the brain of H.M.," he said when we spoke on the phone a few weeks before I met with Corkin. "It was a little bit visionary."

Annese understood how important it was to preserve Henry's brain in as pristine a state as possible. "I remember the unfortunate fate of other illustrious brains," he said. "You know what happened with Einstein's, right?" I did know that bizarre story. When Einstein died at Princeton Hospital in 1955, his body was cremated. Before that happened, however, the hospital pathologist, William Harvey, removed the brain and cut it up into more than two hundred small chunks. He did this without asking the family, although he did persuade them to give their permission after the fact. Harvey doled out a few chunks to researchers, but kept most of them in a couple of Mason jars, swimming in alcohol.

Harvey left Princeton soon thereafter, and ended up in Wichita, Kansas, where journalist Steven Levy tracked him down in 1978. Einstein's pickled brain was still in the Mason jars, under a pile of boxes in Harvey's basement. Eventually, some pieces were acquired by the Mütter Museum of medical oddities, in Philadelphia. You can visit them there, where you can also see a tumor removed from Grover Cleveland's mouth; a slice of tissue from the chest of John Wilkes Booth; the shared liver of the original Siamese twins, Chang and Eng Bunker; and a collection of cysts, tumors, and deformities of all sorts. The research that came out of the few specimens of Einstein's brain that Harvey gave to actual scientists was sparse, and some of it turned out to be questionable.

Annese didn't want this to happen to Henry's brain any more than Suzanne Corkin did. He told the National Science Foundation that he would freeze it, then slice it into more than two thousand thin sheets, each just seventy microns thick. (A micron is equal to .000039 inches; a human hair is about one hundred microns thick.) Scientists would then be able to look at the slices with microscopy and other techniques to learn precisely what Henry's brain looked like, right down to the cellular level.

Annese won the grant, which doesn't surprise David Amaral, director of research at the University of California, Davis's Medical Investigation of Neurodevelopmental Disorders (MIND) Institute, who got involved later in the process. "He has tremendous technical skills in processing human brains," Amaral said. "There aren't very many laboratories in the world that can section an entire human brain in a way that makes it available for any kind of analysis. And Jacopo had set up a laboratory to do that."

Corkin reached Annese while she was still in her car. He immediately booked a flight on the red-eye from San Diego to Boston, where Henry's brain would be extracted at Massachusetts General Hospital. Then she went into her condo, pulled the copy of the flowchart off the kitchen wall, and began calling other members of the recovery team. As soon as Henry had been pronounced dead, meanwhile, attendants at the Bickford Health Care Center, in Windsor Locks, where he had been living, packed ice around his head to slow any deterioration. Then they transferred the body to a hearse, which began the approximately two-hour drive to Boston. The driver was supposed to call Corkin when he got within range, but there was no cell reception inside the building where the brain would be removed, so she waited outside. She was, Corkin writes, "huddled against the frigid Boston weather in a full-length down coat, hood and mittens." At around eight-thirty, "I saw a vehicle rounding the corner, moving hesitantly. I ran toward it, waving my arms over my head. 'I'm Suzanne Corkin! I think you're looking for me!'" As she watched her colleagues transfer Henry's body from the hearse to a gurney, she noticed that he'd been covered with a patchwork quilt. "Somehow," she continues, "I felt comforted by this homey touch."

Before they removed his brain, they put Henry into the MRI for one last, super-detailed look with an especially powerful scanner. The session would last all night. In the morning, Corkin drove out to the airport to pick up Annese, who would be present for the extraction. (They stopped at a Starbucks on the way back, she reports, for a quick espresso.) The removal itself would be performed by Matthew Frosch, Mass General's director of pathology. He would be guided in part by the images from the overnight MRI, which would show him where to cut in order to separate the brain from its surrounding membrane most safely. Frosch was the second person Corkin had called when she got the news earlier that day. "Although Matthew seemed to proceed with complete confidence," Corkin writes, "he later admitted that he had deliberately turned his back to the window [through which Corkin, Annese, and the other observers were watching] so that I would not see how much he was sweating." At one point, Corkin left to call Brenda Milner, who didn't yet know that Henry had died. She evidently took it in stride.

In the end, Frosch got the brain out without damaging it. It went into a bucket of formalin, which transformed it from tofu into something more like clay. (Frosch suspended the brain by a thread tied to the bucket's handle, Corkin says, so that it wouldn't float down and distort itself by flattening against the bottom of the bucket.) From there, Henry's brain went into a solution of formaldehyde. Ten weeks later, Annese, who had gone back to San Diego to get everything ready, returned to Boston, where he transferred Henry's brain to a custom-made cooler packed with ice. He was going to hand-carry it back to San Diego. A PBS camera crew documented the whole thing, including a security team escorting Annese, Corkin, and the brain through Boston's Logan Airport. (Henry and his cooler were exempted from going through the X-ray at the TSA checkpoint.) Then Corkin and Annese hugged, as she left him at the gate to carry his precious carry-on back to California. There's no report on whether Annese's seat mate asked him what he had in the cooler.

The slicing of Henry's brain happened almost precisely a year after he died. Over that time, Corkin said, Annese had left the brain soaking in a mixture of sugar and formaldehyde. The formaldehyde acted as a preservative, while the sugar would keep ice crystals from forming when the brain was frozen prior to slicing. Then Henry's brain had been encased in gelatin to maintain its shape and cooled until the whole thing formed a solid block. Suzanne Corkin flew out for the big event, which was both filmed and Webcast. Annese also invited a reporter from the *New York Times,* as well as what Corkin describes as "several luminaries at his university." The procedure, which lasted fifty-three hours, produced exactly two thousand four hundred and one vertical slices, each seventy microns thick, starting with the front of the brain, right behind where Henry's face would have been, and working backward. It went off without a hitch. "The brain was cut very well," David Amaral said. "There was no problem with that. But . . ." He hesitated, then continued: "From my scientific perspective, there was a lot of unnecessary showmanship." Amaral felt that Jacopo Annese was using his brain-slicing event as a way to attract donations for what Annese was calling the Brain Observatory. On the organization's Web site it says: "The Brain Observatory is committed to maintaining the highest standards in open science, sharing all the images and data that are created in our laboratories with other researchers and the public."

It didn't quite turn out that way, however, at least according to both Amaral and Corkin. A number of the slices were mounted on slides, just as Annese had promised would happen, with the rest preserved for future mounting. But that was evidently pretty much it. Normally, Amaral said, a scientist in Annese's position would develop collaborations with other neuroscientists to study such a valuable specimen. "For example," he said, "my expertise is on the hippocampus, so either I or people like me should have been approached to carry out those studies." Other parts of Henry's brain should have been examined by other experts. But in fact, Amaral said, "nothing happened, nothing happened, nothing happened." Sue Corkin spoke with Amaral and other people in an effort to see what might be done. "We tried to intercede," Amaral said, "and still nothing happened." Annese, Corkin said, "turned out to be a bad collaborator. He basically brought the science to a screeching halt by not living up to his agreements." (Annese did publish a paper in *Nature Communications* in 2013 that described the structural damage to Henry's brain, along with "diffuse pathology in the deep white matter and a small, circumscribed lesion in the left orbitofrontal cortex." According to Corkin, however, who was listed as a coauthor on the paper, this did not amount to a full pathology report.)

Annese rejects these accusations. "It is regrettable," he said by e-mail, "that some colleagues have been led to have this impression about my work with H.M. The question of the neuropath examination, which was only a portion of the overall planned work, has been brought up several times over the years and there's ample evidence of my good will and concrete actions to facilitate the process." At the institutional level, he wrote, "there have been many delays that were beyond my control. In my opinion, difficulties were spurred by a failure to communicate expectations and intents directly, effectively and in a transparent way. This was very frustrating at times, but I did my very best in being very open about my prerogatives throughout." All of the money for cutting and preserving Henry's brain, he said, came from his own grants, without any contribution from MIT or Mass General. He asked for assurances that the Brain Observatory would receive proper credit in any scientific publications, and that it would have a long-term role in the science. "Oddly," he wrote, "I wasn't able to obtain a suitable response. In fact, I personally had very minimal if no response at all after a certain point."

That's not how Corkin saw it. Finally, she said, she simply got fed up with trying to pry information out of Annese. She enlisted administrators at MIT, Mass General, and the University of California, San Diego, to help force the issue. In the end, the higher-ups agreed that Henry's brain should be transferred to another institution—specifically, to Amaral's lab at the University of California, Davis. "It took a very long time to negotiate that," Amaral said, "and the whole process was not facilitated by Dr. Annese."

Annese's lab at UCSD was shut down, and when he and I spoke by phone, before I learned any of this backstory, he told me he was looking for funding to secure a new home for the Brain Observatory. After I'd heard Corkin's and Amaral's accusations, I asked Annese by e-mail why his university agreed to give custody of Henry's brain and other materials over to another lab. He didn't really answer. "The transfer of the collection," he wrote, "without concrete scientific or logistic reasons followed negotiations at the institutional level, well above my jurisdiction. I resigned from UCSD in February of this year because while I respected the fact that UCSD's delegated leadership felt that this solution was in the University's best interest, their decisions were ultimately not aligned with the long-term success of my lab and projects, which are now governed by an independent non-profit organization."

Now that Henry's brain has been relocated, the research is finally getting under way in earnest. Amaral and his colleagues have begun digitizing the slides Annese prepared, creating high-resolution images that will go up on the Web. They've concluded an agreement with someone Amaral calls "a very well respected neuropathologist" to scrutinize the tissue for diseases that might have caused Henry's cognitive decline. And they've begun putting together several consortia of experts who will look at specific parts of the brain to try to understand precisely what happened when William Scoville inadvertently destroyed Henry's memory-making apparatus more than sixty years ago. "We will be able to provide a very definitive description," Amaral said, "of what has been removed, what's intact."

*Chapter 18*

# WHAT DOES THE HIPPOCAMPUS DO?

When Lonni Sue had finished with the fMRI session at Princeton with Nick Turk-Browne, Jiye Kim, and Nayeon Kim, she wasn't done with her scientific obligations for the day. After she'd put her jewelry, her headband, and her fleece vest back on, refilled the vest's pockets with pens and pencils, and taken back possession of her blue tote bag, it was time to trudge upstairs, accompanied by Turk-Browne, her sister, and me, to the fourth floor of the Princeton Neuroscience Institute. She was scheduled to do an entirely different set of tests up there. These were designed to test the conventional wisdom, established early in Milner's and Corkin's work on H.M., that procedural learning—mirror drawing was the classic example—happened somewhere other than the hippocampi and the medial temporal lobes.

"When I was a graduate student, I always thought it was kind of strange that it was interpreted this way—that basically we had this kind of robotic system that worked automatically and unconsciously," Jordan Taylor said. Taylor was the young neuroscientist who greeted us when we stepped off the elevator that morning; he would be conducting the second round of testing. The reason it seemed strange, he said, was that when he was working with normal subjects on procedural-learning tasks, it was clear that the process wasn't entirely unconscious. "They were aware of what I was doing to them. They could tell me kind of what was going on, and so I turned my research focus toward that," he said.

Along with other neuroscientists, Taylor revisited the idea that the hippocampus was irrelevant to procedural learning. When you look at the experiments on H.M., it seems extraordinary that he could learn a new skill such as mirror drawing even though he couldn't form new declarative memories. What's less well known, Taylor said, is that H.M. didn't learn as well as control subjects doing the same tasks. "If it was any other patient group, you would have called him impaired," he said, "but because of his other, more profound impairments, it got spun the other way, right?"

The tests Taylor would be doing on Lonni Sue that morning involved a sort of flight simulator. She sat at a table in front of a screen that displayed an icon representing an airplane and another representing a runway. Her task was to push on a control stick to guide the plane onto the runway. There were two complications. First, on each attempt, which lasted only a couple of seconds, the runway would appear in a different place—straight ahead, straight behind, or to the right or left. Lonni Sue had to figure how to move the lever in order to "land" safely. The second complication was that Taylor and a graduate student named Sam McDougle had programmed the lever so it resisted her movements. If she tried to move it forward, it would push to the right or to the left, and she'd have to compensate with a counterforce of her own in order to land. "It's like a crosswind," she said. "I wish this airport had a windsock!"

Lonni Sue did figure out how to compensate. She showed procedural learning. But like Henry, she didn't do it as well as normal controls. The other salient fact about this experiment, like the ones with H.M. and even the music sight-reading experiments that Landau, McCloskey, and Gregory had done with Lonni Sue, was that none of these truly involved a new skill. "Trace the star while looking in the mirror" is really just a mechanical task. Henry already knew how to hold a pen and follow a line; he simply had to let the link between his visual and motor systems reset itself. Lonni Sue already knew how to read music and how to translate the notes on a page into motor movements; she just had to apply those skills to a composition she hadn't seen before.

But what if she tried to learn an entirely new musical instrument—the trumpet, say, for which you produce the sound not by stroking with a bow but by buzzing your lips into a mouthpiece—tightening and loosening them to make the pitch higher or lower—and form the notes by pressing down three piston-like keys in different patterns? Could she do it? Taylor doesn't know, but he suspects maybe not. In mirror drawing or landing a simulated airplane, which are variations on things amnesics can already do, it's clear that the hippocampus isn't entirely necessary. But it's helpful. In the case of her learning an entirely new skill, such as a new instrument, or maybe a new kind of brush or some other kind of tool she could make drawings with, the hippocampus might be essential. "That's, I think, where I would really like to go," Taylor said. "Come up with some tool that somebody can learn within an hour to get pretty proficient in it—a new tool that they've never used before with totally different physical dynamics, and to see if she can do that."

Taken together, the Princeton and Johns Hopkins experiments with Lonni Sue have helped reinforce a consensus that has been growing over the past few years: that the hippocampus is intimately involved in far more than simply creating and accessing declarative memories. It's not surprising that neuroscientists thought otherwise at first. Lonni Sue and Henry Molaison and Kent Cochrane and

Clive Wearing and E.P. and many other amnesics lost significant chunks of their declarative-memory function when their hippocampi and surrounding tissues were destroyed. But as researchers have looked closer, they've begun to understand that the losses are more widespread than that. In many areas of science, the things you discover are the things you're looking for—and experts in areas of neuroscience other than memory weren't looking to the hippocampus. People who study decision-making, for example, were focused on the frontal lobes, since that's where this process mostly happens. People who studied adaptive learning looked at the object-selective cortex. People who were focused on motor learning looked at the cerebellum and the striatum. There was no reason to think the hippocampus was involved.

But that has begun to change, as neuroscientists have probed more deeply. In Lonni Sue, the loss of her medial temporal lobes has destroyed much of her declarative memory and her ability to form new episodic memories, but it's also affected her ability to do unconscious, statistical learning. It has affected her motor learning—less for simple tasks, but possibly more for sophisticated ones. It has hampered her visual system's ability to adapt to the sight of familiar objects. If you asked the question "What does the hippocampus do?" you'd still have to say it mostly does declarative memory, but you'd also have to note it's involved in varying degrees in other kinds of memory. The full range of that partial involvement is only starting to be explored.

But saying that the function of a brain structure is to do mostly this, but also a little of that, plus a medium amount of something else doesn't really feel like a fundamental explanation, and it doesn't make a whole lot of sense from an evolutionary perspective. "The memory system is really highly conserved across all mammals," Howard Eichenbaum, a neuroscientist at Boston University, said, which is to say that these systems arose early in mammalian evolution, and have persisted in virtually all the descendant species of those remote ancestors. That being the case, Daphna Shohamy, a neuroscientist at Columbia University, said, "this traditional focus on declarative memories doesn't quite fit. It almost seems to ignore the fact that animals have a hippocampus."

To Shohamy, as to many other neuroscientists, this suggests that the question needs to be reframed. "If you zoom out to think about how the brain is built and what it's doing," she said, "and you look at the connectivity between the hippocampus and other parts of the brain, the idea that the hippocampus is just a module for conscious awareness doesn't seem like the most plausible explanation of how the brain is set up." Or, as Ken Norman at Princeton puts it, "It's one thing to say, 'Oh, the hippocampus is important, you know, for our ability to remember specific past events.' It's another thing to really have a satisfying explanation of exactly how a machine with this funny wiring diagram—right?—would have that property."

So if, instead of thinking about what the hippocampus is for, you focus more on what it's uniquely good at doing, things start to make a lot more sense. The hippocampus is in charge of memory, but in a much, much broader way than most of us think about it. It now appears, according to research over the past several decades on both human and animal subjects, that the hippocampus is highly specialized at creating associations among objects, spaces, and experiences. It helps the brain link all of these elements into networks that help us transform the chaotic jumble of sensory impressions that pour into our brains at every minute into a comprehensible whole. Neuroscientists call this linking capability "relational processing," and while it's crucial to our having rich memories of the past, it's also crucial to integrating experience into the present, and for using it to think about the future.

During my visit to the University of Illinois, Neal Cohen gave me an example. Imagine that he and I had just met, and that I'd mentioned I was writing about a patient Nick Turk-Browne was working with at Princeton. Cohen's immediate response might have been to think of the last time he'd seen Nick, and about their conversation, and who else was there. The scenario sounded like my actual encounter with Aline on the streets of Princeton, and the memories it triggered of middle school. That ability to move back and forth between related events, separated in time, is impaired in patients with hippocampal damage, Cohen said. "That's not a deficit someone would come into the clinic and complain about," he said. They'd complain about not remembering what happened yesterday, or about not knowing who their spouse is, or something like that. These deficits are more subtle, but they prevent patients from relating new experiences to old ones, for predicting, based on past experience, the possible outcomes of every decision we make, every moment of the day. Relational processing, Turk-Browne said, "is how you select what actions to perform or what to say or what to think." It doesn't matter whether those decisions are conscious or unconscious.

One of the first hints that relational processing might be what the hippocampus does came out of studies published by John O'Keefe, at University College London, in 1971. By measuring the neural activity of rats moving around an enclosure, O'Keefe and his colleague Jonathan Dostrovsky discovered individual cells that fired as the animals moved through their environment. The neurons, which they called "place cells," evidently formed a mental representation of space—a neural map that recorded the relationships between locations, allowing the rodents to navigate more efficiently with time. The psychologist Edward Tolman had argued as far back as 1930 that rats formed mental maps of their environments, but this was the first evidence of how they did it.

In the early 2000s, May-Britt and Edvard Moser, of the Norwegian University of Science and Technology, found another type of cell in the entorhinal cortex, a medial-temporal-lobe structure that feeds directly into the hippocampus and receives information back from it. Named *grid cells* because they fire off in hexagonal patterns, these keep a record of where the rodents are located within their environment at any given time. These two discoveries earned O'Keefe and the Mosers the 2014 Nobel Prize in Physiology or

Medicine. Later, the Mosers and other neuroscientists went on to find cells in the medial temporal lobes that tell the rats how their heads are oriented within the environment; cells that fire off to signal the rats how far they are from the walls of their enclosures; and cells that keep track of the animals' running speed, which they might use to gauge how far they've traveled from their starting points.

Working together, all of these neural cells evidently serve as a sort of natural GPS system, a form of relational processing that orients the rats in space. The hippocampus also has a set of cells that orient rats in time. Known as (what else?) *time cells,* they were discovered by Howard Eichenbaum. Since memory systems have been conserved during mammalian evolution, humans presumably have them as well.

Even before the Mosers discovered grid cells or Eichenbaum discovered time cells, John O'Keefe and his collaborator Lynn Nadel, of the University of Arizona, had begun to describe the hippocampus not primarily as a memory structure but as a cognitive map of the world. But to read the map, you have to be able to retrieve it from memory—"not," Turk-Browne said, "by consciously thinking, 'Okay, I turn left here.' Sometimes you do that, when you're going to a new place, but most of the time, you're just kind of efficiently moving through the world." It's the hippocampus that lets you do that. "You know about those classic studies of taxi drivers, right?"

He was talking about research by Eleanor Maguire, of London's Institute of Neurology, on the brains of London cabbies. Unlike their counterparts in other major cities, cab drivers in London have to pass an exhaustive test of their geographical knowledge—known simply as "the Knowledge"—before they can get a license. Once they're on the streets, that knowledge is further informed by the dozens of acts of navigation they perform each day, adding a visceral component to their factual understanding of how to get from here to there. When Maguire and her colleagues put London cab drivers into MRI machines and scanned their brains, they found that the cab drivers' hippocampi were significantly enlarged compared with non-cabbies in the part of the organ where spatial information is processed. Then Maguire compared the taxi drivers with London bus drivers, who presumably drive the city more than average citizens but who use fixed routes. They don't really navigate. Once again the taxi drivers stood out. (They did worse, however, on tests that measured how well they learned new spatial information about places other than London; evidently, focusing on knowing one place in extraordinary detail left little room for much else.)

Space and time are two things our brains navigate. So are the worlds of objects, faces, tastes, body sensations, customs, social hierarchies, landscapes, and more. What the hippocampus does, in the metaphor of Princeton's Ken Norman, is to tie together everything we're experiencing, moment by moment. "If you think of the pattern of brain activity during some event as just a bunch of balloons bobbing around, the hippocampus ties all the strings of those balloons together into a little knot." The little knot, he said, is the hippocampal code for that event. Once it's made the knot, then all of that brain activity—sights, sounds, emotions, sensations—is linked. "So now when you tug down on that knot, all the balloons go down, or if you tug on one balloon you can access the other balloons."

These networks of information are stored in other parts of the brain, including the visual cortex, the auditory cortex, and so on. But the hippocampus is what linked them together in the first place—the host at the party, or the balloon man, whatever you prefer. It processes the constant stream of impressions flowing in through our senses and into the brain, and works with the cortex to sort out what's new and what needs updating ("birds fly" updated with "except penguins and ostriches, even though they're also birds"). It establishes relationships between things in the world. "There's been a slow, but sure, change in the consensus," Daphna Shohamy said, "and it's an exciting thing to see."

*Chapter 19*

# LONNI SUE'S WORLD

When Aline talks about Lonni Sue's "puzzle world," she's not simply talking about the puzzles themselves. "When she draws a line," she told me at one point, "it's not just a line. It's a part of a huge structure she's been slowly building for herself over the course of more than seven years."

To help me understand what she means by this, Aline suggests we visit Lonni Sue. She walks me to her sister's room, saying hello to various aides and other staff members along the way. They're all old friends by now. Some of them even took turns standing vigil in the room, just down the hall, where Maggi had lain dying a few months earlier. When we get to the room, Aline tells me to go on in. She'll hang back for a few minutes, just to see how Lonni Sue reacts to me without her sister to distract her.

I walk in, and I'm the one who's distracted. It's a small room, perhaps ten feet wide by twelve feet long, and it's visually overwhelming. The narrower wall to my left, above the neatly made single bed, is plastered with newspaper articles, mostly from local papers, about Lonni Sue, her illness, and her recovery, including several about the art exhibition Barbara Landau helped put together. The longer wall in front of me and the wall to my right are covered with artwork, most of it Lonni Sue's, including a T-shirt she designed for the Cherry Valley International Kite Festival, framed and unframed paintings in many sizes, and a copy of the Princeton Poster. There's also a small black-and-white photograph of Lonni Sue at the age of perhaps thirteen or fourteen, looking intently into the camera, her father standing beside her. "My mother had some things framed and put those up," Aline said before we got there, "and then Lonni Sue did all the rest." When she'd wallpapered the room with artwork as high as she could easily reach, she asked for more pushpins so she could begin work on the out-of-reach area. Aline said no. "I can't do that, because she'll be climbing up on the furniture, and maybe falling."

Along the long wall, below the artwork, is a series of black wire shelving units, forming a long countertop the length of the room. The shelves are full of books, while the top surface holds a line of open boxes filled with colored file folders. In front of the boxes, Lonni Sue has propped up books and pamphlets, many of which she illustrated. The catalog from Maggi's 2004 show, which Lonni Sue read from at the memorial service, is among them. So is the program from the memorial service itself. Lonni Sue put it there as a reminder that their mother is gone. When Lonni Sue notices the program, with Maggi's photograph on the front, she talks about her mother's absence. She'll also talk about it when Aline brings up Maggi's death, of course, and she even brings it up spontaneously. When she does, she sounds wistful, but not grief-stricken. You don't get a sense that the loss has dramatically affected her—although it's impossible to know what's really going on in her mind.

Right in front of me, taking up most of the rest of the room, is

Lonni Sue's work area—two small tables, it looks like, joined to form a makeshift desk. I say "looks like" because the tables themselves are mostly invisible under a portable light-table Lonni Sue uses for tracing. There is also a Scrabble board she's drawn and covered with words (only some of which are actual words), and eight or ten open tissue boxes, each packed with eight small, empty, open-topped blue Cream-O-Land pint milk cartons. Each of these is filled in turn with colored pens and pencils and markers, all carefully organized by type and by color and—in the case of the pencils—by whether they're sharpened, or need sharpening, or have been sharpened so many times that they're too short to draw with. Lonni Sue hates to throw anything away.

She keeps her watercolor brushes in a separate shoe box, organized by size and standing in upright cardboard toilet-paper tubes. She also has several milk cartons lined outside the shoe boxes, each containing a different kind of pencil sharpener. A homemade shelving unit, formed out of milk cartons lying on their sides at the end of the desk, against the wall, holds pads of sticky notes and other odds and ends. Balanced atop the milk-carton shelves she's got open tissue boxes with a forest of colored pencils, point-side up, sticking out of them.

This is Lonni Sue's puzzle world—the environment she's carefully created in service of the life's work she's assigned herself. She's been doing her signature four-page puzzles since late 2010. She hasn't finished any of them, but Aline is convinced that finishing isn't the point. "What we see on the page looks very sparse," she said, "but what we see is not what she sees. She sees all the possibilities that page presents." The puzzles, Aline explained, are a tool, a scaffolding Lonni Sue uses to make connections between letters and words and ideas and the few memories she retains—the essential elements of her internal life. "She tries to capture everything that crosses her mind. What's on the paper exists for as long as that paper is in front of her." It's the same with her odd, disjointed conversations. If Aline reminds Lonni Sue to brush her teeth, for example, her sister might respond that "brush" has the word "rush" in it. "For me, that's a distraction," Aline said, "but in her world, that's the way things are related. Different things matter to her."

Lonni Sue never throws her unfinished puzzles away. That's what the colored folders on the wire shelving in her room are for. She's surrounded by hundreds upon hundreds of partially completed puzzles, plus everything she needs to make thousands more, all of it within easy reach. She didn't come up with this layout all at once. She's rearranged the room a dozen times or more since she moved in, trying to get it exactly right. Barbara Landau actually thinks this might be another aspect of Lonni Sue's creativity that scientists could study. "You can imagine asking questions," she said during a visit to Johns Hopkins, "along the lines of: Here are some materials. How many different ways can you figure out to use these materials to build an environment? She would be way off the scale on that."

Another crucial element of the puzzle world is the ever-present schedule. Without it, Aline has said more than once, Lonni Sue would be floating around in time, with no sense of the past or the future. She still doesn't hold those concepts the way most people do, but at least the schedule has artificially re-created the sense of mental time travel Endel Tulving talked about—the sense whose loss is the most debilitating symptom of amnesia. Lonni Sue isn't consciously aware of this loss, but she must feel it at some level. She clings to her schedule as if it were a life preserver, which in a real sense it is, now that it helps her know when she'll be forced to break away from her puzzles and when she can return.

All of which explains why, when I step through the door of her room, Lonni Sue looks up, beams at me, and says, "Oh, hello! I'm just writing your name on my schedule!" It's already there: Lini made certain of that the night before. But Lonni Sue wants to write it again, just to be sure. "Do you mind? Is that okay?" It's more than okay, I tell her, while I try to take in the posters and drawings and boxes and milk cartons and everything else. Then she says, "I have a lot of things to show you!"

The first thing she wants to show me is a slim, squarish book, an anthology of her artwork printed by a publisher in Japan. It's called *Planet News,* for reasons she no longer remembers (of course). The cover illustration shows a globe with tiny people spaced around its perimeter. Dialogue and thought balloons emerge from their heads with undecipherable script inside, as a crescent moon looks down benignly. It's classic Lonni Sue. Some of the illustrations in the book are *New Yorker* covers, including a stylized beach scene and a picture of shoppers waiting to mail Christmas presents in a line that snakes back and forth in an ever-narrowing series that forms the shape of a Christmas tree (the one that is Amy Goldstein's favorite; it also appears on the cover of a collection of *New Yorker* Christmas covers).

Another illustration, titled "The Glaciers Melt," has a global-warming theme: it shows Manhattan with only the tops of a few tall buildings, plus the tops of the Brooklyn and George Washington bridges, poking out of the rising ocean. Still others come from a book titled *57 Reasons Not to Have a Nuclear War;* they include scenes of people looking at paintings in a museum; opera-goers at a performance; people gazing up at the Rockefeller Center Christmas tree as skaters glide in the foreground; a cat perched cozily on the back of an easy chair while a man reads the newspaper; newlyweds riding a tandem bicycle, he still in a tux and top hat and she in her wedding gown.

As she leafs through the book, she describes each illustration for me. "Oh, I did this one for *The New Yorker,*" she says, pointing. It feels like she's remembering, but she's probably not, since the original source of the drawings is printed at the bottom of each page. She's also proud of the short poems that appear in the book. "I wrote these," she tells me, then reads one aloud:

*Each day is a surprise package*
*to open slowly or fast,*
*the contents scattered*
*and enjoyed with abandon*
*or*

*studied and savored and*
*organized meticulously.*
*Some days—not to be*
*opened at all,*
*but just carried along*
*heavily*
*hoarded up*
*until*
*tomorrow.*

It doesn't occur to her that she's no longer capable of hoarding up the contents of one day to be opened up tomorrow.

Lonni Sue is, as always, relentlessly cheerful. Her drawings are cheerful. Her poems are cheerful. Once, Nick Turk-Browne said, they were running a test in which she had to react to photographs of faces, saying whether she liked them or not. The idea was to test whether she would tend to prefer faces she'd seen on an earlier run to ones she'd never seen. "Liking" in this case would be a proxy for familiarity—a faculty that's still preserved in many medial-temporal-lobe amnesia patients—even though she had no conscious memories of having seen the faces before. Unaccountably, she didn't like any of them, and Maggi explained why. All of the faces had been chosen with deliberately neutral expressions, so test-takers would be able to compare them fairly with one another. But Maggi pointed out that Lonni Sue could see the neutrality as anger or sadness, and she had no stomach for negative emotions. Her alphabet songs had to have happy, uplifting words like "inspiring" and "beautifully" and "creatively." Her drawings were all cats and horses and stars and smiling suns and whimsy.

"Maybe you'd like to show him how you create your puzzles," Aline suggests. She's finally come through the door, now that Lonni Sue and I have gotten started. We've reached the focus of Lonni Sue's existence—the puzzles she has created, entirely on her own, to give meaning to her life.

What comes next is almost impossible for me to follow. She

plucks a sheet of paper out of a box on the table. It's like graph paper, except that Lonni Sue made it herself. She describes the paper, not by the size of the boxes created by the crisscrossing lines, but by the number of boxes that fit across the page and the number down. "Isn't that interesting?" she says, counting the boxes on this sheet. "Twenty by twenty-seven. If you minus two on each side you get eighteen by twenty-five, so you can go five blocks by nine-by-nine blocks which is three-by-three blocks. Let me just . . . so minus two . . ." I'm lost already.

It turns out that she's calculating how many letters she can fit across and down the page. Once she's decided what size grid she wants to use, she lays the graph paper on her light-box, puts another sheet on top of it, and traces out the basic structure for the first page of a new puzzle, using the graph paper as a guide. She takes out a colored marker and starts drawing the boxes that will be filled with letters to frame the page. Sometimes the lines she draws are ruler-straight, but not now. "It's helpful to do wavery lines if you're in a hurry," she explains.

"Because they don't have to be perfect?" I ask.

"Well, they have their own rhythm, which is sort of nice. 'Rhythm' . . . words with a 'y' in them are fun. Have you gone through the alphabet using words with a 'y'?" I hadn't, so we give it a try. She's happy to take the lead. "Aye, bye, cyclone . . ." She's stumped when we get to "f," so Aline suggests "fry." We never get through the alphabet, because Lonni Sue notices the grid in front of her and remembers that she's supposed to be showing me how she makes a puzzle. The theme will be "swimming," she decides.

Aline interrupts again, asking Lonni Sue to tell me why she loves words so much. "Oh, it's so interesting," she says. "Our vocabularies that are in our . . . well, I think the things that we study stretch the span of our verbal knowledge just because of things . . . the words you have to know to do different things. Like music, like, they don't use the word 'pace' so much but 'space,' add 's' to 'p' and you get 'space.' And the tempo of music, the speed of it and how it alters, whether it's brisk or slow, you know, I haven't thought of 'brisk' forever. It's got the word 'risk' in it." Both Aline and I are having a hard time following, but Lonni Sue seems to be making perfect sense to herself. I chime in with the observation that the ancient Greek and Hebrew alphabets start with the equivalent of A and B, just like ours. "That's so *interesting,*" Lonni Sue says. "I wonder if that came from some myth of a god coming and saying, 'A, B, C.'"

Now I ask Lonni Sue who taught her how to draw. "Well, Mummy is an artist. Didn't she do art things with us, Lini? Do you remember what she told us?" She talked about the visual language, Aline reminded her. "It's nice to think about how we can communicate with people in different ways," Lonni Sue says.

I ask her whether she spends a lot of time doing her puzzles. "When I'm allowed to," she says. When isn't she allowed to? "When I sleep. But I work on them in my own way then." Does that mean she dreams about puzzles? "Yes. Now, let's see, I think I have this sorted out," she says, going back to the puzzle. "That's 'sorted,' not 'sordid,'" she explains, just in case I wondered.

I try to engage her on something other than puzzles and words and the alphabet. Aline told me Lonni Sue is interested in astronomy, so I tell her I sometimes write about it—galaxies and stars and planets. "That's fascinating how it's all out there," she says. "I mean, just think of what we are on right now. Something that's swirling around the Sun. We're swirling. We don't seem to go that fast but we really do." Why don't we feel it? I wonder. "Well, we do because we feel gravity. If we took our souls over we'd feel it." And then, "I love space because I used to have two airplanes." Lonni Sue has ventured away from her repertoire of preferred subjects for long enough. She needs to veer back now.

Suddenly she looks at her watch. "I have to go; it's time for lunch. I have just five minutes." I want to know just one more thing, though. I pick up the program from her mother's service, which is now four months in the past. I ask her if she knows what it is. "It's a lovely show or something. We played music at it. That's my mother," she says, pointing to the photograph of Maggi on the front." I ask if she remembers what the show was about. "Yes. She died."

. . .

When Aline first stopped me on the street to tell me the story of what happened to her sister, I had been appalled. The idea of losing my entire history aside from the general biographical facts sounded terrifying. Without my memories—by which I meant my episodic memories, although I didn't think of it this way at the time—I was convinced that my identity would disappear. I would no longer have a self. It seemed as though it would be an awful fate.

But it's clear that Lonni Sue hasn't lost her self. In many ways, she's the same person she was before she became amnesic. She still has her artistic and musical talent, and even if they're not expressed exactly as they were before, they give her enormous pleasure. She still loves to play with words. She's still a workaholic. She's warm, gracious, and hilarious. People still fall in love with her, just as they did before.

As far as her internal life is concerned, it's obviously impossible to say what that might feel like to her, but Michael Graziano, a Princeton neuroscientist who specializes in the phenomenon of consciousness, has thought about it. Graziano hasn't studied Lonni Sue in a formal way. He knows about her because he's married to Sabine Kastner. When the Johnsons came over to the house for Sunday coffee on that late spring day, he had the chance to observe her casually, and, he said, "I'm not sure that Lonni Sue's state of consciousness is as alien as you would think."

It's true that she has major gaps in her knowledge, but people with brain damage tend not to be aware of holes in their awareness. The classic example, he said, is a condition called *hemispatial neglect.* "You get damage to one side of your brain," he said, "typically the right hemisphere, typically this area right above the ear." He indicated the spot on his own head. "And you lose all awareness that there is a left side of space. [The right side of the brain, paradoxically, is aware of things on your left, and vice versa.] It's like *boom*—it's gone and you don't notice." If you ask someone with this kind of injury to describe the room they're in, they'll describe the right half of the room and think they're done. "They don't notice they did anything weird," he said. If you ask them to describe the room they grew up in from memory, the same thing happens. Not only has the left side of the world disappeared; as far as they're concerned it never existed—and they aren't even aware of it. The empty space in Lonni Sue's world has a different character, but the principle, he suspects, is the same. Enormous swaths of memory have simply vanished—as Aline told me once, "a data cassette is missing"—but as far as Lonni Sue is concerned, it never existed at all. Sometimes she'll say, "Why can't I remember that?," as if she knows how a normal memory should operate. Usually, however, she's unaware of her loss. From her perspective, she's the same person she always was.

Another crucial aspect of our self-awareness, Graziano said, is a sense of our bodies. It is, he said, "kind of foundational to our larger psychological sense of self. We have a body schema, and it kind of tells us where the parts of our bodies are and what configuration they're in at any moment—but also, really crucially, what belongs to us and what doesn't." It's possible to trick this sense with the sensory equivalent of an optical illusion: in the so-called rubber-hand illusion, psychologists have subjects sit with their hands resting on a table. The right hand is visible; the left is hidden behind a barrier. There's also an artificial left hand, which the subject can see, in a position where the real one might plausibly rest. A cloth is draped over the left shoulder and down over the table, so while the subject knows that it isn't her real hand, it looks as though it could be. The experimenter strokes her real left hand with a brush, and also strokes the artificial hand, in the same way, at the same time. "If it's set up right," Graziano said, "and the stimulations are really simultaneous, it can be very powerful. Cognitively, you know it's a rubber hand, but you feel like it is real, and it's really strange." There's also the opposite effect, which Oliver Sacks writes about in *The Man Who*

*Mistook His Wife for a Hat.* He describes a man with brain damage who is convinced that his actual leg doesn't belong to him. He's repelled by it, and tries to throw it out of his hospital bed, but since it's attached, he ends up flinging his entire body onto the floor. Lonni Sue is nothing like that. "Her sense of a bodily self distinct from the world—'This belongs to me'—that's all there," Graziano said.

It's true that her attention is almost always very narrowly focused on a few simple things—her puzzles, her schedule, her sister (when Aline is around, that is)—and only on one of them at a time. But that's true for all of us. "When I'm eating a hamburger," Graziano said, "that might be all that's on my mind. Everything else is gone, and because it's all gone I'm not missing it." Or maybe he's thinking about something else and doesn't notice that he's eating the hamburger. We can come out of these states of focus and Lonni Sue probably can't, but again, she doesn't know she's stuck, so her subjective experience isn't that much different from ours. "She feels things. She's aware of things. She's conscious of things. It's just that the things she's conscious of are restricted."

In the abstract, it's natural to think of Lonni Sue's amnesia as a tragedy. When she's in the room with you, it's much more difficult to do so. It's true that conversations go around in circles, and it's disconcerting when she asks you three or four times in a twenty-minute period what your name is, and whether you like to sing alphabet songs. But she's almost invariably charming and full of joy. You can't spend any time with her without bursting into laughter—and she'll be laughing more than you will. "She walks around with her full identity wherever she goes," Aline said. "With what she carries, and with what she says."

When Sabine Kastner and Michael Graziano had the Johnsons over for German coffee hour, their daughter, Sarah, who works as a speech pathologist in New York, happened to be there. She was naturally curious, having never met someone with amnesia before. Afterward, she said to her mother, "If you hadn't told me about

Lonni Sue, I don't think I would necessarily even have noticed that there was something wrong." Kastner agrees. "I think it's quite remarkable," she said. "I think it's exceptional how well she seems to have adapted to this condition, and how fulfilled she seems to be, as a human being."

For those who knew her before, spending time with her can be more difficult. The last time Joe Yacinski and Ron Flemmings came to Princeton, they joined Maggi, Aline, and Lonni Sue for lunch at Maggi's house. It wasn't a comfortable visit for Yacinski. "We had a nice lunch," he said, "but I got a sense that she really would rather have been working on her drawings. She seemed anxious about it."

Flemmings had an entirely different reaction. He and Lonni Sue went off by themselves for a while. "We went to draw," he said. "We drew. Yeah. See, for me, that trip was the best because I was isolated with her, in the studio, drawing and laughing and having fun, and it was joyous. It was great to be back in that space with her. It was wonderful."

# EPILOGUE

Every night, whether or not she's already been out to see Lonni Sue earlier in the day, and no matter how long they've spent together, Aline drives out to get her sister ready for bed. She started doing this a couple of years into recovery because Lonni Sue never remembered to brush her teeth before bed, and ultimately had to have extensive, expensive dental work. At some point, however, Aline realized that her sister probably couldn't go to sleep the way you or I might. "We turn off the light," Aline said, "and then continue thinking for a while, and finally drift off to sleep." Maybe we reflect on the day that just passed, or maybe we think about what's coming tomorrow.

Without her schedule to look at, however, Lonni Sue can't do these things. The thought of her sister just lying there in the dark with nothing going on in her head struck Aline as kind of horrifying. So she began staying after the brushing was finished, to guide Lonni Sue through that transition. Aline switches off the light, and the sisters talk for fifteen or twenty minutes. "We talk about art," she said, "the role of art in culture." This is evidently a favorite subject of Lonni Sue's. "She has such insightful ideas. I never . . . I mean, they show the full wisdom of someone who is mid-sixties, who has had an intense career in art." Once, for example, Lonni Sue came up with the observation that art is a vehicle. What did she mean? Aline asked. "Art is a way of expressing feelings we have," Lonni Sue said, "of expressing ideas, and thoughts. It's a universal language. It's one of the ways we have of communicating."

Or they might talk about the family. "She actually is like a repository of memory for our family," Aline said, and then added, "the general things." That is to say, unsurprisingly, that Lonni Sue isn't a repository for episodic memories. Aline comes up with those, to which Lonni Sue might say, "I can't believe Mum's not here anymore, but she'll always be here, just like Daddy." Aline might ask for particulars about the family, but, she said, "Lonni Sue can't go very far with that. She can see my father's smile or my mother's smile and their creativity, but it's a little vague, although maybe there's more than she can put into words. I'm not quite sure. Anyway," she said, "I might tell some stories, and just having her as an audience, we hold on to the family. It's really very touching."

Or Aline might ask Lonni Sue what she'll be doing tomorrow. The very last thing Lonni Sue does before the light goes out is to check her calendar for the next day. She might remember one or two things, but then she might say, "It's going to be Friday, right? Are we in November?" Sometimes she gets it, other times she might be quite off. "Is there a holiday? Have we had a holiday in November?" She might ask about that kind of thing, trying to orient herself in time. Then she might ask, "Lini, have you ever taken all the letters of the word 'Friday' and mixed them up and see what you can get?" They've gone through this sort of thing a million times, Aline said. "It's fun, though, if you can let yourself have fun with it."

Or she might ask Aline what she did today. "Did you have a good day, Lini? Are you drawing? What are you going to do tomorrow?" And then, just before Aline tiptoes out, she says, "Sleep well. I love you. Sweet dreams." After which they review the schedule one last time. "She says she has to know what to dream about when she goes to sleep," Aline said. She might dream about what's on her schedule, or maybe, Aline thinks, she dreams about flying, or about the alphabet, and all the words she can think of starting with an "A," starting with a "B," starting with a "C." There's no way to know, because the next day, Lonni Sue won't remember.

No one could have imagined when the Johnson girls were young and ambitious, and beginning their careers in art and music, and not particularly close to each other, that they'd find themselves bound so tightly together in the aftermath of a random assault by a common virus. No one could have imagined, either, as Lonni Sue hovered on the edge of death in an upstate New York hospital, that the sisters would end up forming such a loving bond, or that each of them would find such a deep and unexpected sense of purpose—Lonni Sue in her puzzle world and Aline in watching over Lonni Sue. As far as anyone can see into the future, they'll spend every evening like this, with the lights out, Aline soothing her sister gently into sleep. The research will go on: Lonni Sue's remarkable brain still has many secrets to reveal about the mysterious thing we call memory. And every morning, Lonni Sue Johnson will wake up, eager to embrace life and puzzles and the alphabet and Lini. She'll be mostly unaware of all that she has lost. But most of the time, it doesn't seem to matter at all.

# ACKNOWLEDGMENTS

This book would not have been even remotely possible without generous contributions of time and energy from a long list of people. Chief among them were Aline and Margaret Johnson, who spent hours upon hours telling me the story of their remarkable family, about the terrible events at the end of December 2007, and about what happened afterward. They also introduced me to the scientists who are studying Lonni Sue, and to many of her friends—including her fellow artists and pilots—so that I could get the broadest possible sense of a woman who can no longer remember most details of her history. I'm grateful to Lonni Sue Johnson as well. She probably doesn't remember our visits together, and won't recognize me if we meet again, but she welcomed me into her life—a gift that was invaluable to me, and to the story I tell in these pages.

The list of scientists who allowed me into their offices and laboratories and gave up their valuable time to speak with me about Lonni Sue's case, and about the neuroscience of memory in general, is too long to reproduce here. Among them, however, the researchers who continue to study Lonni Sue's memory—primarily Barbara Landau, Michael McCloskey, Emma Gregory, Nicholas Turk-Browne, and Sabine Kastner—were especially helpful. I'm indebted to Ken Norman as well, for allowing me to sit in on his class on memory at Princeton University. The rest of the scientists I spoke with, including the pioneering memory researchers Eric Kandel and Suzanne Corkin (the latter now deceased), are named in the references section. Also too numerous for this space are Lonni Sue's friends and professional acquaintances from throughout her extraordinary life, all of whom were eager to tell me about the woman who made such a powerful and positive impression on them. Rebecca Weil, Buzz Stetson, Robert Landau, and Harry and Ellen Levine stand out especially, but the rest were invariably candid and enormously helpful. Every one of them clearly loved Lonni Sue. Their names, too, are listed in full in the references.

If you find yourself drawn into this story, it's in large part thanks to the many insightful comments I got from those who read the manuscript at various stages along the way, including Joanna Foster, Hannah Lemonick, David LaMotte, Greta Shum, Lauren Feldman, and Isabella Gomes. Nina Rouhani lent her expert eye as a graduate student in neuroscience to the entire manuscript and helped me avoid a number of technical errors (as did many of the other scientists who read smaller sections).

I can't say enough good things about my literary agent, Eleanor Jackson, who helped me turn a run-of-the-mill proposal into something that could catch an editor's eye. I'm also indebted to my editor, Melissa Danaczko, whose thoughtful stewardship helped me tell this story far better than I could possibly have done otherwise, and to her assistant, Margo Shickmanter, whose mastery of the publishing process and whose reassuring style helped keep me on track at several points when I might otherwise have veered off. Anke Steineke was also extremely helpful with the legal nuances—also very reassuring.

Finally, and always, I am indebted to my family, and especially to my wife, Eileen Hohmuth-Lemonick, whose love and support have never faltered. Thank you for taking this journey with me, in both the narrower and the much broader sense.

# AUTHOR'S NOTE

## METHODS

Between July 2013 and January 2016, I conducted interviews, either in person or over the telephone, with members of Lonni Sue Johnson's family, with friends of hers from throughout her life, with the scientists who have studied her and continue to do so, with other scientists who do research on memory and amnesia, and, finally, with Lonni Sue herself. Unless noted otherwise, all quotes attributed to these people come from those interviews. In addition, all of the general information about and descriptions of Lonni Sue's life come from information gathered in those interviews.

My interview subjects are as follows (in alphabetical order, by section):

FAMILY

Aline Johnson
Lonni Sue Johnson
Margaret (Maggi) Johnson

CHILDHOOD FRIENDS

Danae Meray-Horvath
Emily Speagle

OTHER FRIENDS FROM LIFE IN PRINCETON

Curt Carlson
Dudley Carlson
Bob Denby
Robert Landau
Marsha Levin-Roger
Henry Martin (the cartoonist)

FRIENDS FROM LIFE IN NEW YORK CITY

Ron Flemmings
Henry Martin (Lonni Sue's ex-husband)
Joe Yacinski

PILOT FRIENDS

Bob Burke
Nicholas Frirsz
Karen Henriques

COOPERSTOWN FRIENDS

Kay Anichini
Pati Grady
Jim Kevlin
Chris Kjolhede
Ellen Levine
Harry Levine
Carol Little
Pamela Livingston
Lynn Marsh
Deborah Sentochnik (infectious-disease specialist at Bassett Hospital; for privacy reasons, would not confirm whether she knew Lonni Sue personally)
Amy Stetson
Buzz Stetson
Tara Sumner
Ted Sumner
Henry Weil
Rebecca Weil

SCIENTISTS WHO STUDY/HAVE STUDIED LONNI SUE

Emma Gregory
Sabine Kastner
Jiye Kim
Barbara Landau
Mike McCloskey
Joel Ramirez
Nicholas Turk-Browne

OTHER SCIENTISTS WHO WORK ON MEMORY OR OTHER ASPECTS OF BRAIN FUNCTION

David Amaral
Jacopo Annese
Bill Bialek
Neal Cohen
Suzanne Corkin
Howard Eichenbaum
Michael Graziano
Eric Kandel
Elizabeth Loftus
James McGaugh
Morris Moscovitch
Ken Norman
Nina Rouhani
Julia Shaw
Daphna Shohamy
Larry Squire
Jordan Taylor

# NOTES

### CHAPTER 1: A TEXTBOOK CASE

14 One day in 1934: My accounts of Henry Molaison's life, his surgery, and its aftermath are based on Suzanne Corkin's *Permanent Present Tense* (New York: Basic Books, 2013).

16 "This frankly experimental operation": W. B. Scoville and B. Milner, "Loss of Recent Memory After Bilateral Hippocampal Lesions," *Journal of Neurology, Neurosurgery, and Psychiatry* 20, no. 1 (1957): 11–21.

19 Milner and Scoville wrote up the results: Ibid.

### CHAPTER 3: WHERE DOES MEMORY LIVE?

34 Broca, who worked at the medical school: Memoir of Paul Broca, *The Journal of the Anthropological Institute of Great Britain and Ireland* 10 (1881): 242–61.

35 "The man was gross, profane, coarse, and vulgar": Anonymous, "A Most Remarkable Case," reprinted from the *Philadelphia Ledger* in *American Phrenological Journal* 13–14 (1851): 89.

35 Lashley called this idea the *theory of mass action*: Larry Squire and Eric Kandel, *Memory from Mind to Molecules,* 2nd edition (Greenwood Village, Colorado: Roberts & Company, 2009), 9.

35 "I sometimes feel, in reviewing the evidence": K. S. Lashley, "In Search of the Engram," *Symposium of the Society for Experimental Biology* 4 (1950): 477–78.

37 For a week, she and her colleagues: Suzanne Corkin, *Permanent Present Tense* (New York: Basic Books, 2013), 50.

38 "the knowledge of an event": William James, *The Principles of Psychology,* vol. 1 (New York: Henry Holt and Company, 1890), 648.

### CHAPTER 4: PRINCETON

46 "My favorite vision": Lonni Sue Johnson, "An Apron Has Two Strings," *Margaret Kennard Johnson: From Stone to Mesh—Sixty Years* (Exhibition Catalog, Rider University Art Gallery, 2004).

### CHAPTER 5: HOW CELLS REMEMBER

58 The seminal paper Miller and Scoville wrote: W. B. Scoville and B. Milner, "Loss of Recent Memory After Bilateral Hippocampal Lesions," *Journal of Neurology, Neurosurgery, and Psychiatry* 20, no. 1 (1957): 11–21.

59 "another biologist might well have": Eric Kandel, *In Search of Memory* (New York: W. W. Norton, 2006 [paper]), 55.

61 With up to one hundred billion neurons: Larry Squire and Eric Kandel, *Memory from Mind to Molecules,* 2nd edition (Greenwood Village, Colorado: Roberts & Company, 2009), 30.

62 "from the naïve notion of trying": Kandel, *In Search of Memory,* 116.

62 "The news had a powerful impact": Ibid., 117.

65 In the end, they'd mapped out precisely: I. Kupfermann and E. R. Kandel, "Neuronal Controls of a Behavioral Response Mediated by the Abdominal Ganglion of Aplysia," *Science* 164, no. 3881 (1969): 847–50.

65 But it was only theoretical: I. Kupfermann et al., "Neuronal Correlates of Habituation and Dishabituation of the Gill-Withdrawal Reflex in Aplysia," *Science* 16, no. 3296 (1979): 1743–45.

### CHAPTER 7: FLIGHT TO COOPERSTOWN

92 "We turn towards Otsego Lake": Lonni Sue Johnson, "An Aerial Perspective," *The Freeman's Journal,* May 18, 2007.

92 "After all of the rain": Lonni Sue Johnson, "An Aerial Perspective," *The Freeman's Journal,* July 28, 2006.

## CHAPTER 10: SECOND AND THIRD TURNING POINTS

136 Hypergraphia, which may have afflicted: Wikipedia.
145 "It was as if every waking moment": Deborah Wearing, *Forever Today* (London: Doubleday, 2005), 127.
145 "I haven't heard anything, seen anything'": Ibid., 147.
145 "Proper consciousness at 1:19 am": Ibid., 148.
146 "'Look!' he said. 'It's new!'": Ibid., 126.
161 "The results—tabulated, analyzed, and laid out": J. Valtonen et al., "New Learning of Music After Bilateral Medial Temporal Lobe Damage: Evidence from an Amnesic Patient," *Frontiers in Human Neuroscience* 8 (2014): 394.

## CHAPTER 11: PICTURES OF LONNI SUE'S BRAIN

176 But Turk-Browne and his colleagues can see: N. Turk-Browne et al., "Neural Evidence of Statistical Learning: Efficient Detection of Visual Regularities Without Awareness," *Journal of Cognitive Neuroscience* 21, no. 10 (2009): 1934–45.
176 In 2000, however, Princeton became: Brett Tomlinson, "Mind Matters: A New Neuroscience Institute Promises to Fast-Track Princeton Research on the Brain," *Princeton Alumni Weekly,* April 18, 2007 (Web version).

## CHAPTER 12: FALSE MEMORY

182 Simply by changing the phrasing of a question: E. F. Loftus and J. C. Palmer, "Reconstruction of Automobile Destruction: An Example of the Interaction Between Language and Memory," *Journal of Verbal Learning and Verbal Behavior* 13 (1974): 585–89.
186 The kind of research documented in the study: J. Shaw and S. Porter, "Constructing Rich False Memories of Committing Crime," *Psychological Science* 26, no. 3 (2015): 291–301.
186 A series of studies on memories of the 1986 *Challenger*: Eugene Winograd and Ulric Neisser, *Affect and Accuracy in Recall* (Cambridge: Cambridge University Press, 1992), various pages.
186 Those kinds of inaccuracies: W. Hirst et al., "Long-Term Memory for the Terrorist Attack of September 11," *Psychological Science* 14, no. 5 (2003): 455–61.
188 What had happened was that: D. Schiller et al., "Preventing the Return of Fear in Humans Using Reconsolidation Update Mechanisms," *Nature* 463 (2010): 49–53.
188 She and her colleague Giuliana Mazzoni: E. Loftus and G. Mazzoni, "Using Imagination and Personalized Suggestion to Change People," *Behavior Therapy* 29 (1998): 691–706.

CHAPTER 13: CHALLENGING THE CONVENTIONAL WISDOM

191 This experiment showed that it was essential: A. C. Schapiro et al., "The Necessity of the Medial Temporal Lobe for Statistical Learning," *Journal of Cognitive Neuroscience* 8 (2014): 1736–47.

192 What happens at each of the multiple processing areas: D. J. Fellemen and D. C. Van Essen, "Distributed Hierarchical Processing in the Primate Cerebral Cortex," *Cerebral Cortex* 1 (1991): 1–47.

193 A patient known as D.F.: M. A. Goodale et al., "A Neurological Dissociation Between Perceiving Objects and Grasping Them," *Nature* 349 (1999): 154–56.

194 He spent several hours with her: Oliver Sacks, "The Abyss," *The New Yorker,* September 24, 2007.

CHAPTER 15: THE OPPOSITE OF AMNESIA

214 For several years, Price: E. S. Parker et al., "A Case of Unusual Autobiographical Remembering," *Neurocase* 12 (2006): 35–49.

215 "If you try to picture Albert Einstein": FAQ page of Joshua Foer's Web site (joshuafoer.com).

216 "This finding is intriguing," wrote McGaugh: J. McGaugh and A. LePort, "The Discovery of Super Memories," *Scientific American,* February 1, 2014.

CHAPTER 16: OTHER AMNESICS

219 Unlike, say, *50 First Dates*: S. Baxendale, "Memories Aren't Made of This: Amnesia at the Movies," *British Medical Journal* 329, no. 7480 (2004): 1480–83.

220 What happened next was more awful: M. C. Duff et al., "Successful Life Outcome and Management of Real-World Memory Demands Despite Profound Retrograde Amnesia," *Journal of Clinical and Experimental Neuropsychology* 30, no. 8 (2008): 931–45.

223 He was "stuporous" for the next five: R. S. Rosenbaum et al., "The Case of K.C.: Contributions of a Memory-Impaired Person to Memory Theory," *Neuropsychologia* 43 (2005): 989–1021.

224 In fact, Tulving invented the term: Endel Tulving and Wayne Donaldson, *Organization of Memory* (London: Academic Press, 1972), 381–402.

225 "Unidirectionality of time": E. Tulving, "Episodic Memory: From Mind to Brain," *Annual Review of Psychology* 53 (2002): 1–25.

228 If the hippocampus is gone: L. Nadel and M. Moscovitch, "Memory Consolidation, Retrograde Amnesia and the Hippocampal Complex," *Current Opinion in Neurobiology* 7 (2007): 217–27.

CHAPTER 17: THE SAGA OF HENRY'S BRAIN

234 In a groundbreaking study of 678 nuns: D. A. Snowdon, "Healthy Aging and Dementia: Findings from the Nun Study," *Annals of Internal Medicine* 139 (2003): 450–54.

235 "My assistant had laminated": Suzanne Corkin, *Permanent Present Tense* (New York: Basic Books, 2013), 288.

237 "Somehow," she continues: Ibid., 290–91.

238 "Although Matthew seemed to proceed": Ibid., 296.

239 "The Brain Observatory is committed": www.thebrainobservatory.org.

239 Annese did publish a paper: J. Annese et al., "Postmortem Examination of Patient H.M.'s Brain Based on Histological Sectioning and Digital 3D Reconstruction," *Nature Communications* 5, no. 3122 (2013).

CHAPTER 18: WHAT DOES THE HIPPOCAMPUS DO?

247 One of the first hints: J. O'Keefe and J. Dostrovsky, "The Hippocampus as a Spatial Map: Preliminary Evidence from Unit Activity in the Freely-Moving Rat," *Brain Research* 34 (1971): 171–75.

247 Named *grid cells* because: E. Moser et al., "Place Cells, Grid Cells, and the Brain's Spatial Representation System," *Annual Reviews of Neuroscience* 31 (2008): 68–89.

248 The hippocampus also has a set of cells: H. Eichenbaum, "Time Cells in the Hippocampus: A New Dimension for Mapping Memories," *Nature Reviews Neuroscience* 15 (2014): 732–44.

248 Even before the Mosers discovered: John O'Keefe and Lynn Nadel, *The Hippocampus as a Cognitive Map* (Oxford: Clarendon Press, 1978).

248 When Maguire and her colleagues put London cab drivers: E. Maguire et al., "Navigation-Related Structural Change in the Hippocampi of Taxi Drivers," *Proceedings of the National Academy of Sciences* 97, no. 8 (2000): 4398–403.

249 Once again the taxi drivers: E. Maguire et al., "London Taxi and Bus Drivers: A Structural MRI and Neuropsychological Analysis," *Hippocampus* 6 (2006): 1091–1101.

CHAPTER 19: LONNI SUE'S WORLD

254 "Each day is a surprise package": Lonni Sue Johnson, *Planet News* (Tokyo: Kodansha, 1995), pages unnumbered.

# BIBLIOGRAPHY

Baddeley, A., M. Eysenck, and M. Anderson. *Memory.* 2nd edition. London and New York: Psychology Press, 2015.

Boswell, James. *The Life of Samuel Johnson, LL.D.* New York: Alexander V. Blake, 1844.

Corkin, Suzanne. *Permanent Present Tense.* New York: Basic Books, 2013.

Graziano, Michael. *Consciousness and the Social Brain.* London: Oxford University Press, 2015 (paper).

James, William. *The Principles of Psychology.* Vol. 1. New York: Henry Holt and Company, 1890.

Johnson, Lonni Sue. *Planet News.* Tokyo: Kodansha, 1995 (ISBN 978-4062078238).

Kandel, Eric. *In Search of Memory.* New York: W. W. Norton, 2006 (paper).

Meck, Su. *I Forgot to Remember.* New York: Simon & Schuster, 2014.

O'Keefe, John, and Lynn Nadel. *The Hippocampus as a Cognitive Map.* Oxford: Clarendon Press, 1978.

Squire, Larry, and Eric Kandel. *Memory from Mind to Molecules.* 2nd edition. Greenwood Village, Colorado: Roberts & Company, 2009.

Tulving, Endel, and Wayne Donaldson. *Organization of Memory.* London: Academic Press, 1972.

Wearing, Deborah. *Forever Today.* London: Doubleday, 2005.

Winograd, Eugene, and Ulric Neisser. *Affect and Accuracy in Recall.* Cambridge: Cambridge University Press, 1992.

Grateful acknowledgment is made to the following for permission to reprint previously published material:

Aline Johnson: Excerpt from "Life" by Lonni Sue Johnson, published in *Planet News* (Kodansha Ltd., 1995). Reprinted by permission of Aline Johnson.

*The Freeman's Journal:* Excerpts from "An Aerial Perspective" by Lonni Sue Johnson, dated May 18, 2007, and July 28, 2006. Reprinted by permission of *The Freeman's Journal.*

Rider University Art Gallery: Excerpt from "An Apron Has Two Strings" by Lonni Sue Johnson, from the art exhibit entitled "From Stone to Mesh—Sixty Years" by Margaret Kennard Johnson in 2004. Reprinted by permission of Rider University Art Gallery.

ABOUT THE AUTHOR

Michael D. Lemonick is the opinion editor at *Scientific American*. He has written more than fifty *Time* magazine cover stories on science, and has been a contributor to *National Geographic*, *The New Yorker*, and other publications. This is his seventh book.